T0282602

CAMBRIDGE LIBRARY COLLECTION

Books of enduring scholarly value

Darwin

Two hundred years after his birth and 150 years after the publication of 'On the Origin of Species', Charles Darwin and his theories are still the focus of worldwide attention. This series offers not only works by Darwin, but also the writings of his mentors in Cambridge and elsewhere, and a survey of the impassioned scientific, philosophical and theological debates sparked by his 'dangerous idea'.

The Malay Archipelago

Alfred Russel Wallace (1823–1913) was a British naturalist who is best remembered as the co-discoverer, with Darwin, of natural selection. His extensive fieldwork and advocacy of the theory of evolution led to him being considered one of the nineteenth century's foremost biologists. These volumes, first published in 1869, contain Wallace's acclaimed and highly influential account of extensive fieldwork he undertook in modern Indonesia, Malaysia and New Guinea between 1854 and 1862. Wallace describes his travels around the island groups, depicting the unusual animals and insects he encountered and providing ethnographic descriptions of the indigenous peoples. Wallace's analysis of biogeographic patterns in Indonesia (later termed the Wallace Line) profoundly influenced contemporary and later evolutionary and geological thought concerning both Indonesia and other areas of the world where similar patterns were found. Volume 2 covers the Molucca Islands and New Guinea.

Cambridge University Press has long been a pioneer in the reissuing of out-of-print titles from its own backlist, producing digital reprints of books that are still sought after by scholars and students but could not be reprinted economically using traditional technology. The Cambridge Library Collection extends this activity to a wider range of books which are still of importance to researchers and professionals, either for the source material they contain, or as landmarks in the history of their academic discipline.

Drawing from the world-renowned collections in the Cambridge University Library, and guided by the advice of experts in each subject area, Cambridge University Press is using state-of-the-art scanning machines in its own Printing House to capture the content of each book selected for inclusion. The files are processed to give a consistently clear, crisp image, and the books finished to the high quality standard for which the Press is recognised around the world. The latest print-on-demand technology ensures that the books will remain available indefinitely, and that orders for single or multiple copies can quickly be supplied.

The Cambridge Library Collection will bring back to life books of enduring scholarly value (including out-of-copyright works originally issued by other publishers) across a wide range of disciplines in the humanities and social sciences and in science and technology.

The Malay Archipelago

*The Land of the Orang-utan,
and the Bird of Paradise.
A Narrative of Travel, with Studies
of Man and Nature*

VOLUME 2

ALFRED RUSSEL WALLACE

CAMBRIDGE
UNIVERSITY PRESS

CAMBRIDGE UNIVERSITY PRESS

Cambridge, New York, Melbourne, Madrid, Cape Town, Singapore,
São Paolo, Delhi, Dubai, Tokyo, Mexico City

Published in the United States of America by Cambridge University Press, New York

www.cambridge.org
Information on this title: www.cambridge.org/9781108022828

© in this compilation Cambridge University Press 2010

This edition first published 1869
This digitally printed version 2010

ISBN 978-1-108-02282-8 Paperback

THE

MALAY ARCHIPELAGO.

VOL. II.

NATIVES OF ARU SHOOTING THE GREAT BIRD OF PARADISE.

THE

MALAY ARCHIPELAGO:

THE LAND OF THE

ORANG-UTAN, AND THE BIRD OF PARADISE.

A NARRATIVE OF TRAVEL,

WITH STUDIES OF MAN AND NATURE.

BY

ALFRED RUSSEL WALLACE,

AUTHOR OF

"TRAVELS ON THE AMAZON AND RIO NEGRO," "PALM TREES OF THE AMAZON," ETC.

IN TWO VOLS.—VOL. II.

London:

MACMILLAN AND CO.

1869.

LONDON :
R. CLAY, SONS, AND TAYLOR, PRINTERS,
BREAD STREET HILL.

MALAY ARCHIPELAGO.

CHAPTER XXI.

THE MOLUCCAS—TERNATE.

ON the morning of the 8th of January, 1858, I arrived at
Ternate, the fourth of a row of fine conical volcanic
islands which skirt the west coast of the large and almost
unknown island of Gilolo. The largest and most perfectly
conical mountain is Tidore, which is over four thousand
feet high—Ternate being very nearly the same height, but
with a more rounded and irregular summit. The town
of Ternate is concealed from view till we enter between
the two islands, when it is discovered stretching along
the shore at the very base of the mountain. Its
situation is fine, and there are grand views on every
side. Close opposite is the rugged promontory and beautiful volcanic cone of Tidore; to the east is the long
mountainous coast of Gilolo, terminated towards the north

by a group of three lofty volcanic peaks, while immediately behind the town rises the huge mountain, sloping
easily at first and covered with thick groves of fruit trees,
but soon becoming steeper, and furrowed with deep gullies.
Almost to the summit, whence issue perpetually faint
wreaths of smoke, it is clothed with vegetation, and looks
calm and beautiful, although beneath are hidden fires
which occasionally burst forth in lava-streams, but more
frequently make their existence known by the earthquakes
which have many times devastated the town.

I brought letters of introduction to Mr. Duivenboden, a
native of Ternate, of an ancient Dutch family, but who
was, educated in England, and speaks our language perfectly. He was a very rich man, owned half the town,
possessed many ships, and above a hundred slaves. He
was moreover, well educated, and fond of literature and
science—a phenomenon in these regions. He was generally known as the king of Ternate, from his large property and great influence with the native Rajahs and their
subjects. Through his assistance I obtained a house,
rather ruinous, but well adapted to my purpose, being
close to the town, yet with a free outlet to the country and
the mountain. A few needful repairs were soon made,
some bamboo furniture and other necessaries obtained, and
after a visit to the Resident and Police Magistrate I found
myself an inhabitant of the earthquake-tortured island of

Ternate, and able to look about me and lay down the plan of my campaign for the ensuing year. I retained this house for three years, as I found it very convenient to have a place to return to after my voyages to the various islands of the Moluccas and New Guinea, where I could pack my collections, recruit my health, and make preparations for future journeys. To avoid repetitions, I will in this chapter combine what notes I have about Ternate.

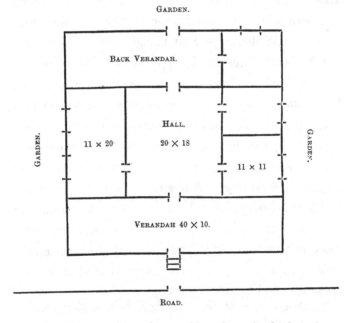

A description of my house (the plan of which is here shown) will enable the reader to understand a very

common mode of building in these islands. There is of
course only one floor. The walls are of stone up to three
feet high ; on this are strong squared posts supporting the
roof, everywhere except in the verandah filled in with the
leaf-stems of the sago-palm, fitted neatly in wooden
framing. The floor is of stucco, and the ceilings are like
the walls. The house is forty feet square, consists of four
rooms, a hall, and two verandahs, and is surrounded by a
wilderness of fruit trees. A deep well supplied me with
pure cold water, a great luxury in this climate. Five
minutes' walk down the road brought me to the market
and the beach, while in the opposite direction there were
no more European houses between me and the mountain.
In this house I spent many happy days. Returning to it
after a three or four months' absence in some uncivilized
region, I enjoyed the unwonted luxuries of milk and fresh
bread, and regular supplies of fish and eggs, meat and
vegetables, which were often sorely needed to restore my
health and energy. I had ample space and convenience
for unpacking, sorting, and arranging my treasures, and I
had delightful walks in the suburbs of the town, or up the
lower slopes of the mountain, when I desired a little
exercise, or had time for collecting.

The lower part of the mountain, behind the town of
Ternate, is almost entirely covered with a forest of fruit
trees, and during the season hundreds of men and women,

boys and girls, go up every day to bring down the ripe fruit. Durians and Mangoes, two of the very finest tropical fruits, are in greater abundance at Ternate than I have ever seen them, and some of the latter are of a quality not inferior to any in the world. Lansats and Mangūstans are also abundant, but these do not ripen till a little later. Above the fruit trees there is a belt of clearings and cultivated grounds, which creep up the mountain to a height of between two and three thousand feet, above which is virgin forest, reaching nearly to the summit, which on the side next the town is covered with a high reedy grass. On the further side it is more elevated, of a bare and desolate aspect, with a slight depression marking the position of the crater. From this part descends a black scoriaceous tract, very rugged, and covered with a scanty vegetation of scattered bushes as far down as the sea. This is the lava of the great eruption near a century ago, and is called by the natives " batu-angas " (burnt rock).

Just below my house is the fort, built by the Portuguese, below which is an open space to the beach, and beyond this the native town extends for about a mile to the north-east. About the centre of it is the palace of the Sultan, now a large untidy, half-ruinous building of stone. This chief is pensioned by the Dutch Government, but retains the sovereignty over the native population of the island, and of the northern part of Gilolo. The sultans of Ternate and Tidore were once celebrated through the East for their power and regal magnificence. When Drake visited Ternate in 1579, the Portuguese had been driven out of the island, although they still had a settlement at Tidore. He gives a glowing account of the Sultan : " The King had a very rich canopy with embossings of gold borne over him, and was guarded with twelve lances. From the waist to the ground was all cloth of gold, and that very rich; in the attire of his head were finely wreathed in, diverse rings of plaited gold, of an inch or more in breadth, which made a fair and princely show, somewhat resembling a crown in form; about his neck he had a chain of perfect gold, the links very great and one fold double; on his left hand was a diamond, an emerald, a ruby, and a turky ; on his right hand in one ring a big and perfect turky, and in another ring many diamonds of a smaller size."

All this glitter of barbaric gold was the produce of the spice trade, of which the Sultans kept the monopoly, and by which they became wealthy. Ternate, with the small islands in a line south of it, as far as Batchian, constitute the ancient Moluccas, the native country of the clove, as well as the only part in which it was cultivated. Nutmegs and mace were procured from the natives of New Guinea and the adjacent islands, where they grew wild ; and the profits on spice cargoes were so enormous, that the European traders were glad to give gold and jewels, and the finest manufactures of Europe or of India, in exchange. When the Dutch established their influence in these seas, and relieved the native princes from their Portuguese oppressors, they saw that the easiest way to repay themselves would be to get this spice trade into their own hands. For this purpose they adopted the wise principle of concentrating the culture of these valuable products in those spots only of which they could have complete control. To do this effectually it was necessary to abolish the culture and trade in all other places, which they succeeded in doing by treaty with the native rulers. These agreed to have all the spice trees in their possessions destroyed. They gave up large though fluctuating revenues, but they gained in return a fixed subsidy, freedom from the constant attacks and harsh oppressions of the Portuguese, and a continuance of their regal power and exclusive authority over their own subjects, which is maintained in all the islands except Ternate to this day.

It is no doubt supposed by most Englishmen, who have been accustomed to look upon this act of the Dutch with vague horror, as something utterly unprincipled and barbarous, that the native population suffered grievously by this destruction of such valuable property. But it is certain that this was not the case. The Sultans kept this lucrative trade entirely in their own hands as a rigid

monopoly, and they would take care not to give their subjects more than would amount to their usual wages, while
they would surely exact as large a quantity of spice as they
could possibly obtain.   Drake and other early voyagers
always seem to have purchased their spice-cargoes from the
Sultans and Rajahs, and not from the cultivators.   Now
the absorption of so much labour in the cultivation of this
one product must necessarily have raised the price of food
and other necessaries; and when it was abolished, more
rice would be grown, more sago made, more fish caught,
and more tortoise-shell, rattan, gum-dammer, and other
valuable products of the seas and the forests would be obtained.   I believe, therefore, that this abolition of the spice
trade in the Moluccas was actually beneficial to the inhabitants, and that it was an act both wise in itself and
morally and politically justifiable.

In the selection of the places in which to carry on the
cultivation, the Dutch were not altogether fortunate or
wise.   Banda was chosen for nutmegs, and was eminently
successful, since it continues to this day to produce a large
supply of this spice, and to yield a considerable revenue.
Amboyna was fixed upon for establishing the clove cultivation; but the soil and climate, although apparently very
similar to that of its native islands, is not favourable, and
for some years the Government have actually been paying
to the cultivators a higher rate than they could purchase

cloves elsewhere, owing to a great fall in the price since the rate of payment was fixed for a term of years by the Dutch Government, and which rate is still most honourably paid.

In walking about the suburbs of Ternate, we find everywhere the ruins of massive stone and brick buildings, gateways and arches, showing at once the superior wealth of the ancient town and the destructive effects of earthquakes. It was during my second stay in the town, after my return from New Guinea, that I first felt an earthquake. It was a very slight one, scarcely more than has been felt in this country, but occurring in a place that had been many times destroyed by them it was rather more exciting. I had just awoke at gun-fire (5 A.M.), when suddenly the thatch began to rustle and shake as if an army of cats were galloping over it, and immediately afterwards my bed shook too, so that for an instant I imagined myself back in New Guinea, in my fragile house, which shook when an old cock went to roost on the ridge ; but remembering that I was now on a solid earthen floor, I said to myself, "Why, it's an earthquake," and lay still in the pleasing expectation of another shock; but none came, and this was the only earthquake I ever felt in Ternate.

The last great one was in February 1840, when almost every house in the place was destroyed. It began about

midnight on the Chinese New Year's festival, at which
time every one stays up nearly all night feasting at the
Chinamen's houses and seeing the processions. This prevented any lives being lost, as every one ran out of
doors at the first shock, which was not very severe. The
second, a few minutes afterwards, threw down a great
many houses, and others, which continued all night and
part of the next day, completed the devastation. The line
of disturbance was very narrow, so that the native town a
mile to the east scarcely suffered at all. The wave passed
from north to south, through the islands of Tidore and
Makian, and terminated in Batchian, where it was not felt
till four the following afternoon, thus taking no less than
sixteen hours to travel a hundred miles, or about six miles
an hour. It is singular that on this occasion there was no
rushing up of the tide, or other commotion of the sea, as is
usually the case during great earthquakes.

The people of Ternate are of three well-marked races :
the Ternate Malays, the Orang Sirani, and the Dutch.
The first are an intrusive Malay race somewhat allied to
the Macassar people, who settled in the country at a very
early epoch, drove out the indigenes, who were no doubt
the same as those of the adjacent mainland of Gilolo, and
established a monarchy. They perhaps obtained many of
their wives from the natives, which will account for the
extraordinary language they speak—in some respects closely

allied to that of the natives of Gilolo, while it contains
much that points to a Malayan origin. To most of these
people the Malay language is quite unintelligible, although
such as are engaged in trade are obliged to acquire it.
" Orang Sirani," or Nazarenes, is the name given by the
Malays to the Christian descendants of the Portuguese,
who resemble those of Amboyna, and, like them, speak
only Malay. There are also a number of Chinese merchants, many of them natives of the place, a few Arabs,
and a number of half-breeds between all these races and
native women. Besides these there are some Papuan
slaves, and a few natives of other islands settled here,
making up a motley and very puzzling population, till
inquiry and observation have shown the distinct origin of
its component parts.

Soon after my first arrival in Ternate I went to the
island of Gilolo, accompanied by two sons of Mr. Duivenboden, and by a young Chinaman, a brother of my land-
lord, who lent us the boat and crew. These latter were
all slaves, mostly Papuans, and at starting I saw something
of the relation of master and slave in this part of the
world. The crew had been ordered to be ready at three
in the morning, instead of which none appeared till five,
we having all been kept waiting in the dark and cold
for two hours. When at length they came they were
scolded by their master, but only in a bantering manner,

and laughed and joked with him in reply. Then, just as we
were starting, one of the strongest men refused to go at all,
and his master had to beg and persuade him to go, and
only succeeded by assuring him that I would give him
something; so with this promise, and knowing that there
would be plenty to eat and drink and little to do, the black
gentleman was induced to favour us with his company and
assistance. In three hours' rowing and sailing we reached
our destination, Sedingole, where there is a house belonging to the Sultan of Tidore, who sometimes goes there
hunting. It was a dirty ruinous shed, with no furniture
but a few bamboo bedsteads. On taking a walk into the
country, I saw at once that it was no place for me. For
many miles extends a plain covered with coarse high grass,
thickly dotted here and there with trees, the forest country
cnly commencing at the hills a good way in the interior.
Such a place would produce few birds and no insects, and
we therefore arranged to stay only two days, and then go
on to Dodinga, at the narrow central isthmus of Gilolo,
whence my friends would return to Ternate. We amused
ourselves shooting parrots, lories, and pigeons, and trying to
shoot deer, of which we saw plenty, but could not get one;
and our crew went out fishing with a net, so we did not
want for provisions. When the time came for us to continue our journey, a fresh difficulty presented itself, for our
gentlemen slaves refused in a body to go with us, saying

very determinedly that they would return to Ternate. So their masters were obliged to submit, and I was left behind to get to Dodinga as I could. Luckily I succeeded in hiring a small boat, which took me there the same night, with my two men and my baggage.

Two or three years after this, and about the same length of time before I left the East, the Dutch emancipated all their slaves, paying their owners a small compensation. No ill results followed. Owing to the amicable relations which had always existed between them and their masters, due no doubt in part to the Government having long accorded them legal rights and protection against cruelty and ill-usage, many continued in the same service, and after a little temporary difficulty in some cases, almost all returned to work either for their old or for new masters. The Government took the very proper step of placing every emancipated slave under the surveillance of the police-magistrate. They were obliged to show that they were working for a living, and had some honestly-acquired means of existence. All who could not do so were placed upon public works at low wages, and thus were kept from the temptation to peculation or other crimes, which the excitement of newly-acquired freedom, and disinclination to labour, might have led them into.

CHAPTER XXII.

GILOLO.

(MARCH AND SEPTEMBER 1858.)

I MADE but few and comparatively short visits to this large and little known island, but obtained a considerable knowledge of its natural history by sending first my boy Ali, and then my assistant, Charles Allen, who stayed two or three months each in the northern peninsula, and brought me back large collections of birds and insects. In this chapter I propose to give a sketch of the parts which I myself visited. My first stay was at Dodinga, situated at the head of a deep bay exactly opposite Ternate, and a short distance up a little stream which penetrates a few miles inland. The village is a small one, and is completely shut in by low hills.

As soon as I arrived, I applied to the head man of the village for a house to live in, but all were occupied, and there was much difficulty in finding one. In the meantime I unloaded my baggage on the beach and made some tea, and afterwards discovered a small hut which the

owner was willing to vacate if I would pay him five
guilders for a month's rent. As this was something less
than the fee-simple value of the dwelling, I agreed to
give it him for the privilege of immediate occupation, only
stipulating that he was to make the roof water-tight.
This he agreed to do, and came every day to talk and
look at me; and when I each time insisted upon his
immediately mending the roof according to contract, all
the answer I could get was, " Ea nanti," (Yes, wait a little.)
However, when I threatened to· deduct a quarter guilder
from the rent for every day it was not done, and a guilder
extra if any of my things were wetted, he condescended to
work for half an hour, which did all that was absolutely
necessary.

On the top of a bank, of about a hundred feet ascent from
the water, stands the very small but substantial fort erected
by the Portuguese. Its battlements and turrets have long
since been overthrown by earthquakes, by which its massive structure has also been rent; but it cannot well be
thrown down, being a solid mass of stonework, forming a
platform about ten feet high, and perhaps forty feet square.
It is approached by narrow steps under an archway, and
is now surmounted by a row of thatched hovels, in which
live the small garrison, consisting of a Dutch corporal and
four Javanese soldiers, the sole representatives of the
Netherlands Government in the island. The village is

occupied entirely by Ternate men. The true indigenes of
Gilolo, "Alfuros" as they are here called, live on the
eastern coast, or in the interior of the northern peninsula.
The distance across the isthmus at this place is only two
miles, and there is a good path, along which rice and sago
are brought from the eastern villages. The whole isthmus
is very rugged, though not high, being a succession of little
abrupt hills and valleys, with angular masses of limestone
rock everywhere projecting, and often almost blocking up
the pathway. Most of it is virgin forest, very luxuriant
and picturesque, and at this time having abundance of
large scarlet Ixoras in flower, which made it exceptionally
gay. I got some very nice insects here, though, owing to
illness most of the time, my collection was a small one;
and my boy Ali shot me a pair of one of the most beautiful
birds of the East, Pitta gigas, a large ground-thrush, whose
plumage of velvety black above is relieved by a breast
of pure white, shoulders of azure blue, and belly of vivid
crimson. It has very long and strong legs, and hops about
with such activity in the dense tangled forest, bristling
with rocks, as to make it very difficult to shoot.

In September 1858, after my return from New Guinea,
I went to stay some time at the village of Djilolo, situated
in a bay on the northern peninsula. Here I obtained a
house through the kindness of the Resident of Ternate,
who sent orders to prepare one for me. The first walk into

the unexplored forests of a new locality is a moment of intense interest to the naturalist, as it is almost sure to furnish him with something curious or hitherto unknown. The first thing I saw here was a flock of small parroquets, of which I shot a pair, and was pleased to find a most beautiful little long-tailed bird, ornamented with green, red, and blue colours, and quite new to me. It was a variety of the Charmosyna placentis, one of the smallest and most elegant of the brush-tongued lories. My hunters soon shot me several other fine birds, and I myself found a specimen of the rare and beautiful day-flying moth, Cocytia d'Urvillei.

The village of Djilolo was formerly the chief residence of the Sultans of Ternate, till about eighty years ago, when at the request of the Dutch they removed to their present abode. The place was then no doubt much more populous, as is indicated by the wide extent of cleared land in the neighbourhood, now covered with coarse high grass, very disagreeable to walk through, and utterly barren to the naturalist. A few days' exploring showed me that only some small patches of forest remained for miles round, and the result was a scarcity of insects and a very limited variety of birds, which obliged me to change my locality. There was another village called Sahoe, to which there was a road of about twelve miles overland, and this had been recommended to me as a good place for birds, and as possessing a large population both of Mahometans and Alfuros, which latter race I much wished to see. I set off one morning to examine this place myself, expecting to pass through some extent of forest on my way. In this however I was much disappointed, as the whole road lies through grass and scrubby thickets, and it was only after reaching the village cf Sahoe that some high forest land was perceived stretching towards the mountains to the north of it. About half-way we had to pass a deep river on a bamboo raft, which almost sunk beneath us. This stream was said to rise a long way off to the northward.

Although Sahoe did not at all appear what I expected, I determined to give it a trial, and a few days afterwards obtained a boat to carry my things by sea while I walked overland. A large house on the beach belonging to the Sultan was given me. It stood alone, and was quite open on every side, so that little privacy could be had, but as I only intended to stay a short time I made it do. A very few days dispelled all hopes I might have entertained of making good collections in this place. Nothing was to be found in every direction but interminable tracts of reedy grass, eight or ten feet high, traversed by narrow paths, often almost impassable. Here and there were clumps of fruit trees, patches of low wood, and abundance of plantations and rice grounds, all of which are, in tropical regions, a very desert for the entomologist. The virgin forest that I was in search of, existed only on the summits and on the steep rocky sides of the mountains a long way off, and in inaccessible situations. In the suburbs of the village I found a fair number of bees and wasps, and some small but interesting beetles. Two or three new birds were obtained by my hunters, and by incessant inquiries and promises I succeeded in getting the natives to bring me some land shells, among which was a very fine and handsome one, Helix pyrostoma. I was, however, completely wasting my time here compared with what I might be doing in a good locality, and after a week returned to Ternate, quite disappointed with my first attempts at collecting in Gilolo.

In the country round about Sahoe, and in the interior, there is a large population of indigenes, numbers of whom came daily into the village, bringing their produce for sale, while others were engaged as labourers by the Chinese and Ternate traders. A careful examination convinced me that these people are radically distinct from all the Malay races. Their stature and their features, as well as their disposition and habits, are almost the same as those of the Papuans; their hair is semi-Papuan—neither straight, smooth, and glossy, like all true Malays', nor so frizzly and woolly as the perfect Papuan type, but always crisp, waved, and rough, such as often occurs among the true

c 2

Papuans, but never among the Malays. Their colour alone is often exactly that of the Malay, or even lighter. Of course there has been intermixture, and there occur occasionally individuals which it is difficult to classify; but in most cases the large, somewhat aquiline nose, with elongated apex, the tall stature, the waved hair, the bearded face, and hairy body, as well as the less reserved manner and louder voice, unmistakeably proclaim the Papuan type. Here then I had discovered the exact boundary line between the Malay and Papuan races, and at a spot where no other writer had expected it. I was very much pleased at this determination, as it gave me a clue to one of the most difficult problems in Ethnology, and enabled me in many other places to separate the two races, and to unravel their intermixtures.

On my return from Waigiou in 1860, I stayed some days on the southern extremity of Gilolo; but, beyond seeing something more of its structure and general character, obtained very little additional information. It is only in the northern peninsula that there are any indigenes, the whole of the rest of the island, with Batchian and the other islands westward, being exclusively inhabited by Malay tribes, allied to those of Ternate and Tidore. This would seem to indicate that the Alfuros were a comparatively recent immigration, and that they had come from the orth or east, perhaps from some of the islands of the Pacific. It is otherwise difficult to understand how so many fertile districts should possess no true indigenes.

Gilolo, or Halmaheira as it is called by the Malays and Dutch, seems to have been recently modified by upheaval and subsidence. In 1673, a mountain is said to have been upheaved at Gamokonora on the northern peninsula. All the parts that I have seen have either been volcanic or coralline, and along the coast there are fringing coral reefs very dangerous to navigation. At the same time, the character of its natural history proves it to be a rather ancient land, since it possesses a number of animals peculiar to itself or common to the small islands around it, but almost always distinct from those of New Guinea on the east, of Ceram on the south, and of Celebes and the Sula islands on the west.

The island of Morty, close to the north-eastern extremity of Gilolo, was visited by my assistant Charles Allen, as well as by Dr. Bernstein; and the collections obtained there present some curious differences from those of the main island. About fifty-six species of land-birds are known to inhabit this island, and of these a kingfisher (Tanysiptera doris), a honeysucker (Tropidorhynchus fuscicapillus), and a large crow-like starling (Lycocorax morotensis), are quite distinct from allied species found in Gilolo. The island is coralline and sandy, and we must

therefore believe it to have been separated from Gilolo
at a somewhat remote epoch; while we learn from its
natural history that an arm of the sea twenty-five miles
wide serves to limit the range even of birds of considerable powers of flight.

CHAPTER XXIII.

(OCTOBER 1858.)

ON returning to Ternate from Sahoe, I at once began
making preparations for a journey to Batchian, an
island which I had been constantly recommended to visit
since I had arrived in this part of the Moluccas. After all
was ready I found that I should have to hire a boat, as
no opportunity of obtaining a passage presented itself. I
accordingly went into the native town, and could only find
two boats for hire, one much larger than I required, and
the other far smaller than I wished. I chose the smaller
one, chiefly because it would not cost me one-third as
much as the larger one, and also because in a coasting
voyage a small vessel can be more easily managed, and
more readily got into a place of safety during violent
gales, than a large one. I took with me my Bornean lad
Ali, who was now very useful to me ; Lahagi, a native
of Ternate, a very good steady man, and a fair shooter,
who had been with me to New Guinea ; Lahi, a native of

Gilolo, who could speak Malay, as woodcutter and general assistant; and Garo, a boy who was to act as cook. As the boat was so small that we had hardly room to stow ourselves away when all my stores were on board, I only took one other man named Latchi, as pilot. He was a Papuan slave, a tall, strong black fellow, but very civil and careful. The boat I had hired from a Chinaman named Lau Keng Tong, for five guilders a month.

We started on the morning of October 9th, but had not got a hundred yards from land, when a strong head wind sprung up, against which we could not row, so we crept along shore to below the town, and waited till the turn of the tide should enable us to cross over to the coast of Tidore. About three in the afternoon we got off, and found that our boat sailed well, and would keep pretty close to the wind. We got on a good way before the wind fell and we had to take to our oars again. We landed on a nice sandy beach to cook our suppers, just as the sun set behind the rugged volcanic hills, to the south of the great cone of Tidore, and soon after beheld the planet Venus shining in the twilight with the brilliancy of a new moon, and casting a very distinct shadow. We left again a little before seven, and as we got out from the shadow of the mountain I observed a bright light over one part of the ridge, and soon after, what seemed a fire of remarkable whiteness on the very summit of the hill. I called the

attention of my men to it, and they too thought it merely
a fire ; but a few minutes afterwards, as we got farther off
shore, the light rose clear up above the ridge of the hill,
and some faint clouds clearing away from it, discovered
the magnificent comet which was at the same time
astonishing all Europe. The nucleus presented to the
naked eye a distinct disc of brilliant white light, from
which the tail rose at an angle of about 30° or 35° with
the horizon, curving slightly downwards, and terminating
in a broad brush of faint light, the curvature of which
diminished till it was nearly straight at the end. The
portion of the tail next the comet appeared three or four
times as bright as the most luminous portion of the
milky way, and what struck me as a singular feature was
that its upper margin, from the nucleus to very near the
extremity, was clearly and almost sharply defined, while
the lower side gradually shaded off into obscurity.
Directly it rose above the ridge of the hill, I said to my
men, "See, it's not a fire, it's a bintang ber-ekor " (" tailed-
star," the Malay idiom for a comet).   "So it is," said they ;
and all declared that they had often heard tell of such,
but had never seen one till now.   I had no telescope
with me, nor any instrument at hand, but I estimated the
length of the tail at about 20°, and the width, towards the
extremity, about 4° or 5°.

The whole of the next day we were obliged to stop near

the village of Tidore, owing to a strong wind right in our
teeth. The country was all cultivated, and I in vain
searched for any insects worth capturing. One of my men
went out to shoot, but returned home without a single bird.
At sunset, the wind having dropped, we quitted Tidore,
and reached the next island, Mareh, where we stayed
till morning. The comet was again visible, but not nearly
so brilliant, being partly obscured by clouds, and dimmed
by the light of the new moon. We then rowed across to the
island of Motir, which is so surrounded with coral-reefs
that it is dangerous to approach. These are perfectly flat,
and are only covered at high water, ending in craggy
vertical walls of coral in very deep water. When there is a
little wind, it is dangerous to come near these rocks; but
luckily it was quite smooth, so we moored to their edge,
while the men crawled over the reef to the land, to make
a fire and cook our dinner—the boat having no accommodation for more than heating water for my morning and
evening coffee. We then rowed along the edge of the reef
to the end of the island, and were glad to get a nice
westerly breeze, which carried us over the strait to the
island of Makian, where we arrived about 8 P.M. The
sky was quite clear, and though the moon shone brightly,
the comet appeared with quite as much splendour as
when we first saw it.

The coasts of these small islands are very different

according to their geological formation. The volcanoes, active or extinct, have steep black beaches of volcanic sand, or are fringed with rugged masses of lava and basalt. Coral is generally absent, occurring only in small patches in quiet bays, and rarely or never forming reefs. Ternate, Tidore, and Makian belong to this class. Islands of volcanic origin, not themselves volcanoes, but which have been probably recently upraised, are generally more or less completely surrounded by fringing reefs of coral, and have beaches of shining white coral sand. Their coasts present volcanic conglomerates, basalt, and in some places a foundation of stratified rocks, with patches of upraised coral. Mareh and Motir are of this character, the outline of the latter giving it the appearance of having been a true volcano, and it is said by Forrest to have thrown out stones in 1778. The next day (Oct. 12th), we coasted along the island of Makian, which consists of a single grand volcano. It was now quiescent, but about two centuries ago (in 1646) there was a terrible eruption, which blew up the whole top of the mountain, leaving the truncated jagged summit and vast gloomy crater valley which at this time distinguished it. It was said to have been as lofty as Tidore before this catastrophe.[1]

---

[1] Soon after I left the Archipelago, on the 29th of December, 1862, another eruption of this mountain suddenly took place, which caused great devastation in the island. All the villages and crops were de-

I stayed some time at a place where I saw a new
clearing on a very steep part of the mountain, and obtained a few interesting insects.   In the evening we went
on to the extreme southern point, to be ready to pass across
the fifteen-mile strait to the island of Kaióa.   At five
the next morning we started, but the wind, which had
hitherto been westerly, now got to the south and south-
west, and we had to row almost all the way with a burning sun overhead.   As we approached land a fine breeze
sprang up, and we went along at a great pace ; yet after an
hour we were no nearer, and found we were in a violent
current carrying us out to sea.   At length we overcame it, and got on shore just as the sun set, having been
exactly thirteen hours coming fifteen miles.   We landed
on a beach of hard coralline rock, with rugged cliffs of the
same, resembling those of the Ké Islands (Chap. XXIX.)
It was accompanied by a brilliancy and luxuriance of the
vegetation, very like what I had observed at those islands,
which so much pleased me that I resolved to stay a few
days at the chief village, and see if their animal productions were correspondingly interesting.   While searching
for a secure anchorage for the night we again saw the

stroyed, and numbers of the inhabitants killed.   The sand and ashes fell
so thick that the crops were partially destroyed fifty miles off, at Ternate,
where it was so dark the following day that lamps had to be lighted at
noon.   For the position of this and the adjacent islands, see the map in
Chapter XXXVII.

comet, still apparently as brilliant as at first, but the tail had now risen to a higher angle.

October 14*th.*—All this day we coasted along the Kaióa Islands, which have much the appearance and outline of Ké on a small scale, with the addition of flat swampy tracts along shore, and outlying coral reefs. Contrary winds and currents had prevented our taking the proper course to the west of them, and we had to go by a circuitous route round the southern extremity of one island, often having to go far out to sea on account of coral reefs. On trying to pass a channel through one of these reefs we were grounded, and all had to get out into the water, which in this shallow strait had been so heated by the sun as to be disagreeably warm, and drag our vessel a considerable distance among weeds and sponges, corals and prickly corallines. It was late at night when we reached the little village harbour, and we were all pretty well knocked up by hard work, and having had nothing but very brackish water to drink all day—the best we could find at our last stopping-place. There was a house close to the shore, built for the use of the Resident of Ternate when he made his official visits, but now occupied by several native travelling merchants, among whom I found a place to sleep.

The next morning early I went to the village to find the "Kapala," or head man. I informed him that I wanted

to stay a few days in the house at the landing, and begged
him to have it made ready for me.   He was very civil,
and came down at once to get it cleared, when we found
that the traders had already left, on hearing that I required
it.   There were no doors to it, so I obtained the loan of
a couple of hurdles to keep out dogs and other animals.
The land here was evidently sinking rapidly, as shown by
the number of trees standing in salt water dead and dying.
After breakfast I started for a walk to the forest-covered
hill above the village, with a couple of boys as guides.
It was exceedingly hot and dry, no rain having fallen for
two months.   When we reached an elevation of about two
hundred feet, the coralline rock which fringes the shore
was succeeded by a hard crystalline rock, a kind of meta-
morphic sandstone.   This would indicate that there had
been a recent elevation of more than two hundred feet,
which had still more recently changed into a movement
of subsidence.   The hill was very rugged, but among
dry sticks and fallen trees I found some good insects,
mostly of forms and species I was already acquainted
with from Ternate and Gilolo.   Finding no good paths I
returned, and explored the lower ground eastward of the
village, passing through a long range of · plantain and
tobacco grounds, encumbered with felled and burnt logs,
on which I found quantities of beetles of the family
Buprestidæ of six different species, one of which was new

to me. I then reached a path in the swampy forest where I hoped to find some butterflies, but was disappointed. Being now pretty well exhausted by the intense heat, I thought it wise to return and reserve further exploration for the next day.

When I sat down in the afternoon to arrange my insects, the house was surrounded by men, women, and children, lost in amazement at my unaccountable proceedings; and when, after pinning out the specimens, I proceeded to write the name of the place on small circular tickets, and attach one to each, even the old Kapala, the Mahometan priest, and some Malay traders could not repress signs of astonishment. If they had known a little more about the ways and opinions of white men, they would probably have looked upon me as a fool or a madman, but in their ignorance they accepted my operations as worthy of all respect, although utterly beyond their comprehension.

The next day (October 16th) I went beyond the swamp, and found a place where a new clearing was being made in the virgin forest. It was a long and hot walk, and the search among the fallen trunks and branches was very fatiguing, but I was rewarded by obtaining about seventy distinct species of beetles, of which at least a dozen were new to me, and many others rare and interesting. I have never in my life seen beetles so abundant as they were on this spot. Some dozen species of good-sized golden

Buprestidæ, green rose-chafers (Lomaptera), and long-horned weevils (Anthribidæ), were so abundant that they rose up in swarms as I walked along, filling the air with a loud buzzing hum. Along with these, several fine Longicorns were almost equally common, forming such an assemblage as for once to realize that idea of tropical luxuriance which one obtains by looking over the drawers of a well-filled cabinet. On the under sides of the trunks clung numbers of smaller or more sluggish Longicorns, while on the branches at the edge of the clearing others could be detected sitting with outstretched antennæ ready to take flight at the least alarm. It was a glorious spot, and one which will always live in my memory as exhibiting the insect-life of the tropics in unexampled luxuriance. For the three following days I continued to visit this locality, adding each time many new species to my collection—the following notes of which may be interesting to entomo logists. October 15th, 33 species of beetles; 16th, 70 species; 17th, 47 species; 18th, 40 species; 19th, 56 species—in all about a hundred species, of which forty were new to me. There were forty-four species of Longi-corns among them, and on the last day I took twenty-eight species of Longicorns, of which five were new to me.

My boys were less fortunate in shooting. The only birds at all common were the great red parrot (Eclectus grandis), found in most of the Moluccas, a crow, and a

Megapodius, or mound-maker. A few of the pretty racquet-tailed kingfishers were also obtained, but in very poor plumage. They proved, however, to be of a different species from those found in the other islands, and come nearest to the bird originally described by Linnæus under the name of Alcedo dea, and which came from Ternate. This would indicate that the small chain of islands parallel to Gilolo have a few peculiar species in common, a fact which certainly occurs in insects.

The people of Kaióa interested me much. They are evidently a mixed race, having Malay and Papuan affinities, and are allied to the peoples of Ternate and of Gilolo. They possess a peculiar language, somewhat resembling those of the surrounding islands, but quite distinct. They are now Mahometans, and are subject to Ternate. The only fruits seen here were papaws and pine-apples, the rocky soil and dry climate being unfavourable. Rice, maize, and plantains flourish well, except that they suffer from occasional dry seasons like the present one. There is a little cotton grown, from which the women weave sarongs (Malay petticoats). There is only one well of good water on the islands, situated close to the landing-place, to which all the inhabitants come for drinking water. The men are good boat-builders, and they make a regular trade of it and seem to be very well off.

After five days at Kaióa we continued our journey, and

soon got among the narrow straits and islands which lead
down to the town of Batchian. In the evening we stayed
at a settlement of Galéla men. These are natives of a
district in the extreme north of Gilolo, and are great
wanderers over this part of the Archipelago. They build
large and roomy praus with outriggers, and settle on any
coast or island they take a fancy for. They hunt deer and
wild pig, drying the meat; they catch turtle and tripang;
they cut down the forest and plant rice or maize, and are
altogether remarkably energetic and industrious. They
are very fine people, of light complexion, tall, and with
Papuan features, coming nearer to the drawings and
descriptions of the true Polynesians of Tahiti and Owyhee
than any I have seen.

During this voyage I had several times had an opportunity of seeing my men get fire by friction. A sharp-
edged piece of bamboo is rubbed across the convex surface
of another piece, on which a small notch is first cut. The
rubbing is slow at first and gradually quicker, till it
becomes very rapid, and the fine powder rubbed off ignites
and falls through the hole which the rubbing has cut in
the bamboo. This is done with great quickness and certainty. The Ternate people use bamboo in another way.
They strike its flinty surface with a bit of broken china,
and produce a spark, which they catch in some kind of
tinder.

On the evening of October 21st we reached our destination, having been twelve days on the voyage. It had been fine weather all the time, and, although very hot, I had enjoyed myself exceedingly, and had besides obtained some experience in boat work among islands and coral reefs, which enabled me afterwards to undertake much longer voyages of the same kind. The village or town of Batchian is situated at the head of a wide and deep bay, where a low isthmus connects the northern and southern mountainous parts of the island. To the south is a fine range of mountains, and I had noticed at several of our landing-places that the geological formation of the island was very different from those around it. Whenever rock was visible it was either sandstone in thin layers, dipping south, or a pebbly conglomerate. Sometimes there was a little coralline limestone, but no volcanic rocks. The forest had a dense luxuriance and loftiness seldom found on the dry and porous lavas and raised coral reefs of Ternate and Gilolo ; and hoping for a corresponding richness in the birds and insects, it was with much satisfaction and with considerable expectation that I began my explorations in the hitherto unknown island of Batchian.

CHAPTER XXIV.

BATCHIAN.

I LANDED opposite the house kept for the use of the Resident of Ternate, and was met by a respectable middle-aged Malay, who told me he was Secretary to the Sultan, and would receive the official letter with which I had been provided. On giving it him, he at once informed me I might have the use of the official residence which was empty. I soon got my things on shore, but on looking about me found that the house would never do to stay long in. There was no water except at a considerable distance, and one of my men would be almost entirely occupied getting water and firewood, and I should myself have to walk all through the village every day to the forest, and live almost in public, a thing I much dislike. The rooms were all boarded, and had ceilings, which are a great nuisance, as there are no means of hanging anything up except by driving nails, and not half the conveniences of a native bamboo and thatch cottage. I accordingly inquired for a house outside of the village on the road to the coal mines, and was informed by the Secretary that there was a small one belonging to the Sultan, and that he would go with me early next morning to see it.

We had to pass one large river, by a rude but substantial bridge, and to wade through another fine pebbly stream of clear water, just beyond which the little hut was situated. It was very small, not raised on posts, but with the earth for a floor, and was built almost entirely of the leaf-stems of the sago-palm, called here "gaba-gaba." Across the river behind rose a forest-clad bank, and a good road close in front of the house led through cultivated grounds to the forest about half a mile on, and thence to the coal mines four miles further. These advantages at once decided me, and I told the Secretary I would be very glad to occupy the house. I therefore sent my two men immediately to buy "ataps" (palm-leaf thatch) to repair the roof, and the next day, with the assistance of eight of the Sultan's men, got all my stores and furniture carried up and pretty comfortably arranged. A rough bamboo bedstead was soon constructed, and a table made of boards which I had brought with me, fixed under the window. Two bamboo chairs, an easy cane chair, and hanging shelves suspended with insulating oil cups, so as to be safe from ants, completed my furnishing arrangements.

In the afternoon succeeding my arrival, the Secretary

accompanied me to visit the Sultan. We were kept waiting a few minutes in an outer gate-house, and then ushered
to the door of a rude, half-fortified whitewashed house. A
small table and three chairs were placed in a large outer
corridor, and an old dirty-faced man with grey hair and a
grimy beard, dressed in a speckled blue cotton jacket and
loose red trousers, came forward, shook hands, and asked
me to be seated. After a quarter of an hour's conversation
on my pursuits, in which his Majesty seemed to take great
interest, tea and cakes—of rather better quality than usual
on such occasions—were brought in. I thanked him for
the house, and offered to show him my collections, which
he promised to come and look at. He then asked me to
teach him to take views—to make maps—to get him a
small gun from England, and a milch-goat from Bengal;
all of which requests I evaded as skilfully as I was able,
and we parted very good friends. He seemed a sensible
old man, and lamented the small population of the island,
which he assured me was rich in many valuable minerals,
including gold; but there were not people enough to look
after them and work them. I described to him the great
rush of population on the discovery of the Australian
gold mines, and the huge nuggets found there, with which
he was much interested, and exclaimed, " Oh! if we had
but people like that, my country would be quite as rich !"

The morning after I had got into my new house, I sent

my boys out to shoot, and went myself to explore the road
to the coal mines. In less than half a mile it entered the
virgin forest, at a place where some magnificent trees
formed a kind of natural avenue. The first part was flat
and swampy, but it soon rose a little, and ran alongside
the fine stream which passed behind my house, and
which here rushed and gurgled over a rocky or pebbly
bed, sometimes leaving wide sandbanks on its margins,
and at other places flowing between high banks crowned
with a varied and magnificent forest vegetation. After
about two miles, the valley narrowed, and the road was
carried along the steep hill-side which rose abruptly from
the water's edge. In some places the rock had been cut
away, but its surface was already covered with elegant
ferns and creepers. Gigantic tree-ferns were abundant,
and the whole forest had an air of luxuriance and rich
variety which it never attains in the dry volcanic soil to
which I had been lately accustomed. A little further the
road passed to the other side of the valley by a. bridge
across the stream at a place where a great mass of rock in
the middle offered an excellent support for it, and two miles
more of most picturesque and interesting road brought me
to the mining establishment.

This is situated in a large open space, at a spot where
two tributaries fall into the main stream. Several forest-
paths and new clearings offered fine collecting grounds,

and I captured some new and interesting insects; but as
it was getting late I had to reserve a more thorough
exploration for future occasions. Coal had been discovered
here some years before, and the road was made in order to
bring down a sufficient quantity for a fair trial on the
Dutch steamers. The quality, however, was not thought
sufficiently good, and the mines were abandoned. Quite
recently, works had been commenced in another spot, in
hopes of finding a better vein. There were about eighty
men employed, chiefly convicts; but this was far too
small a number for mining operations in such a country,
where the mere keeping a few miles of road in repair
requires the constant work of several men. If coal of
sufficiently good quality should be found, a tramroad
would be made, and would be very easily worked, owing
to the regular descent of the valley.

Just as I got home I overtook Ali returning from
shooting with some birds hanging from his belt. He
seemed much pleased, and said, " Look here, sir, what a
curious bird," holding out what at first completely puzzled
me. I saw a bird with a mass of splendid green feathers
on its breast, elongated into two glittering tufts; but, what
I could not understand was a pair of long white feathers,
which stuck straight out from each shoulder. Ali assured
me that the bird stuck them out this way itself, when
fluttering its wings, and that they had remained so without

WALLACE'S STANDARD WING, MALE AND FEMALE.

his touching them. I now saw that I had got a great prize, no less than a completely new form of the Bird of Paradise, differing most remarkably from every other known bird. The general plumage is very sober, being a pure ashy olive, with a purplish tinge on the back ; the crown of the head is beautifully glossed with pale metallic violet, and the feathers of the front extend as much over the beak as in most of the family. The neck and breast are scaled with fine metallic green, and the feathers on the lower part are elongated on each side, so as to form a two-pointed gorget, which can be folded beneath the wings, or partially erected and spread out in the same way as the side plumes of most of the birds of paradise. The four long white plumes which give the bird its altogether unique character, spring from little tubercles close to the upper edge of the shoulder or bend of the wing ; they are narrow, gently curved, and equally webbed on both sides, of a pure creamy white colour. They are about six inches long, equalling the wing, and can be raised at right angles to it, or laid along the body at the pleasure of the bird. The bill is horn colour, the legs yellow, and the iris pale olive. This striking novelty has been named by Mr. G. R. Gray of the British Museum, Semioptera Wallacei, or "Wallace's Standard wing."

A few days later I obtained an exceedingly beautiful new butterfly, allied to the fine blue Papilio Ulysses, but differing from it in the colour being of a more intense tint, and in having a row of blue stripes around the margin of the lower wings. This good beginning was, however, rather deceptive, and I soon found that insects, and especially butterflies, were somewhat scarce, and birds in far less variety than I had anticipated. Several of the fine Moluccan species were however obtained. The handsome red lory with green wings and a yellow spot in the back (Lorius garrulus), was not uncommon. When the Jambu, or rose apple (Eugenia sp.), was in flower in the village, flocks of the little lorikeet (Charmosyna placentis), already met with in Gilolo, came to feed upon the nectar, and I obtained as many specimens as I desired. Another beautiful bird of the parrot tribe was the Geoffroyus cyanicollis, a green parrot with a red bill and head, which colour shaded on the crown into azure blue, and thence into verditer blue and the green of the back. Two large and handsome fruit pigeons, with metallic green, ashy, and rufous plumage, were not uncommon; and I was rewarded by finding a splendid deep blue roller (Eurystomus azureus), a lovely golden-capped sunbird (Nectarinea auriceps), and a fine racquet-tailed kingfisher (Tanysiptera isis), all of which were entirely new to ornithologists. Of insects I obtained a considerable number of interesting beetles, including many fine longicorns, among which was the largest and handsomest species of the genus Glenea yet

discovered. Among butterflies the beautiful little Danis
sebæ was abundant, making the forests gay with its delicate wings of white and the richest metallic blue; while
showy Papilios, and pretty Pieridæ, and dark, rich Euplæas,
many of them new, furnished a constant source of interest
and pleasing occupation.

The island of Batchian possesses no really indigenous
inhabitants, the interior being altogether uninhabited, and
there are only a few small villages on various parts of the
coast; yet I found here four distinct races, which would
wofully mislead an ethnological traveller unable to obtain
information as to their origin. First there are the Batchian
Malays, probably the earliest colonists, differing very little
from those of Ternate. Their language, however, seems to
have more of the Papuan element, with a mixture of pure
Malay, showing that the settlement is one of stragglers
of various races, although now sufficiently homogeneous.
Then there are the "Orang Sirani," as at Ternate and
Amboyna. Many of these have the Portuguese physiognomy strikingly preserved, but combined with a skin generally darker than the Malays. Some national customs are
retained, and the Malay, which is their only language,
contains a large number of Portuguese words and idioms.
The third race consists of the Galela men from the north
of Gilolo, a singular people, whom I have already described;
and the fourth is a colony from Tomŏré, in the eastern

peninsula of Celebes. These people were brought here at their own request a few years ago, to avoid extermination by another tribe. They have a very light complexion, open Tartar physiognomy, low stature, and a language of the Bugis type. They are an industrious agricultural people, and supply the town with vegetables. They make a good deal of bark cloth, similar to the tapa of the Polynesians, by cutting down the proper trees and taking off large cylinders of bark, which is beaten with mallets till it separates from the wood. It is then soaked, and so continuously and regularly beaten out that it becomes as thin and as tough as parchment. In this form it is much used for wrappers for clothes ; and they also make jackets of it, sewn neatly together and stained with the juice of another kind of bark, which gives it a dark red colour and renders it nearly waterproof.

Here are four very distinct kinds of people who may all be seen any day in and about the town of Batchian. Now if we suppose a traveller ignorant of Malay, picking up a word or two here and there of the " Batchian language," and noting down the " physical and moral peculiarities, manners, and customs of the Batchian people "—(for there are travellers who do all this in four-and-twenty hours)— what an accurate and instructive chapter we should have ! what transitions would be pointed out, what theories of the origin of races would be developed ! while the next traveller might flatly contradict every statement and arrive at exactly opposite conclusions.

Soon after I arrived here the Dutch Government introduced a new copper coinage of *cents* instead of *doits* (the 100th instead of the 120th part of a guilder), and all the old coins were ordered to be sent to Ternate to be changed. I sent a bag containing 6,000 doits, and duly received the new money by return of the boat. When Ali went to bring it, however, the captain required a written order; so I waited to send again the next day, and it was lucky I did so, for that night my house was entered, all my boxes carried out and ransacked, and the various articles left on the road about twenty yards off, where we found them at five in the morning, when, on getting up and finding the house empty, we rushed out to discover tracks of the thieves. Not being able to find the copper money which they thought I had just received, they decamped, taking nothing but a few yards of cotton cloth and a black coat and trousers, which latter were picked up a few days afterwards hidden in the grass. There was no doubt whatever who were the thieves. Convicts are employed to guard the Government stores when the boat arrives from Ternate. Two of them watch all night, and often take the opportunity to roam about and commit robberies.

The next day I received my money, and secured it well in a strong box fastened under my bed. I took out five or six hundred cents for daily expenses, and put them in a small japanned box, which always stood upon my table. In the afternoon I went for a short walk, and on my return this box and my keys, which I had carelessly left on the table, were gone. Two of my boys were in the house, but had heard nothing. I immediately gave information of the two robberies to the Director at the mines and to the Commandant at the fort, and got for answer, that if I caught the thief in the act I might shoot him. By inquiry in the village, we afterwards found that one of the convicts who was on duty at the Government rice-store in the village had quitted his guard, was seen to pass over the bridge towards my house, was seen again within two hundred yards of my house, and on returning over the bridge into the village carried something under his arm, carefully covered with his sarong. My box was stolen between the hours he was seen going and returning, and it was so small as to be easily carried in the way described. This seemed pretty clear circumstantial evidence. I accused the man and brought the witnesses to the Commandant. The man was examined, and confessed having gone to the river close to my house to bathe; but said he had gone no further, having climbed up a cocoa-nut tree and brought home two nuts, which he had covered over, *because he was ashamed to be seen carrying them!* This explanation was thought satisfactory, and he was acquitted. I lost my

cash and my box, a seal I much valued, with other small
articles, and all my keys—the severest loss by far. Luckily
my large cash-box was left locked, but so were others
which I required to open immediately. There was, however, a very clever blacksmith employed to do ironwork
for the mines, and he picked my locks for me when I
required them, and in a few days made me new keys, which
I used all the time I was abroad.

Towards the end of November the wet season set in, and
we had daily and almost incessant rains, with only about
one or two hours' sunshine in the morning. The flat parts
of the forest became flooded, the roads filled with mud,
and insects and birds were scarcer than ever. On
December 13th, in the afternoon, we had a sharp earthquake shock, which made the house and furniture shake
and rattle for five minutes, and the trees and shrubs wave
as if a gust of wind had passed over them. About the
middle of December I removed to the village, in order
more easily to explore the district to the west of it, and to
be near the sea when I wished to return to Ternate. I
obtained the use of a good-sized house in the Campong
Sirani (or Christian village), and at Christmas and the
New Year had to endure the incessant gun-firing, drum-
beating, and fiddling of the inhabitants.

These people are very fond of music and dancing, and it
would astonish a European to visit one of their assemblies.

We enter a gloomy palm-leaf hut, in which two or three
very dim lamps barely render darkness visible. The floor
is of black sandy earth, the roof hid in a smoky impene-
trable blackness; two or three benches stand against the
walls, and the orchestra consists of a fiddle, a fife, a drum,
and a triangle. There is plenty of company, consisting of
young men and women, all very neatly dressed in white and
black—a true Portuguese habit. Quadrilles, waltzes, polkas,
and mazurkas are danced with great vigour and much
skill. The refreshments are muddy coffee and a few sweet-
meats. Dancing is kept up for hours, and all is conducted
with much decorum and propriety. A party of this kind
meets about once a week, the principal inhabitants taking
it by turns, and all who please come in without much
ceremony.

It is astonishing how little these people have altered
in three hundred years, although in that time they
have changed their language and lost all knowledge of
their own nationality. They are still in manners and
appearance almost pure Portuguese, very similar to those
with whom I had become acquainted on the banks of the
Amazon. They live very poorly as regards their house
and furniture, but preserve a semi-European dress, and
have almost all full suits of black for Sundays. They are
nominally Protestants, but Sunday evening is their grand
day for music and dancing. The men are often good

hunters ; and two or three times a week, deer or wild pigs
are brought to the village, which, with fish and fowls,
enables them to live well. They are almost the only
people in the Archipelago who eat the great fruit-eating
bats called by us " flying foxes." These ugly creatures are
considered a great delicacy, and are much sought after.
At about the beginning of the year they come in large
flocks to eat fruit, and congregate during the day on some
small islands in the bay, hanging by thousands on the
trees, especially on dead ones. They can then be easily
caught or knocked down with sticks, and are brought
home by baskets-full. They require to be carefully prepared, as the skin and fur has a rank and powerful foxy
odour ; but they are generally cooked with abundance of
spices and condiments, and are really very good eating,
something like hare. The Orang Sirani are good cooks,
having a much greater variety of savoury dishes than the
Malays. Here, they live chiefly on sago as bread, with
a little rice occasionally, and abundance of vegetables and
fruit.

It is a curious fact that everywhere in the East where
the Portuguese have mixed with the native races they
have become darker in colour than either of the parent
stocks. This is the case almost always with these " Orang
Sirani " in the Moluccas, and with the Portuguese of
Malacca. The reverse is the case in South America, where

the mixture of the Portuguese or Brazilian with the Indian produces the " Mameluco," who is not unfrequently lighter than either parent, and always lighter than the Indian. The women at Batchian, although generally fairer than the men, are coarse in features, and very far inferior in beauty to the mixed Dutch-Malay girls, or even to many pure Malays.

The part of the village in which I resided was a grove of cocoa-nut trees, and at night, when the dead leaves were sometimes collected together and burnt, the effect was most magnificent—the tall stems, the fine crowns of foliage, and the immense fruit-clusters, being brilliantly illuminated against a dark sky, and appearing like a fairy palace supported on a hundred columns, and groined over with leafy arches. The cocoa-nut tree, when well grown, is certainly the prince of palms both for beauty and utility.

During my very first walk into the forest at Batchian, I had seen sitting on a leaf out of reach, an immense butterfly of a dark colour marked with white and yellow spots. I could not capture it as it flew away high up into the forest, but I at once saw that it was a female of a new species of Ornithoptera or "bird-winged butterfly," the pride of the Eastern tropics. I was very anxious to get it and to find the male, which in this genus is always of extreme beauty. During the two succeeding months I only saw it once again, and shortly afterwards I saw the male flying high in the air at the mining village. I had begun to despair of ever getting a specimen, as it seemed so rare and wild ; till one day, about the beginning of January, I found a beautiful shrub with large white leafy bracts and yellow flowers, a species of Mussænda, and saw one of these noble insects hovering over it, but it was too quick for me, and flew away. The next day I went again to the same shrub and succeeded in catching a female, and the day after a fine male. I found it to be as I had expected, a perfectly new and most magnificent species, and one of the most gorgeously coloured butterflies in the world. Fine specimens of the male are more than seven inches across the wings, which are velvety black and fiery orange, the latter colour replacing the green of the allied species. The beauty and brilliancy of this insect are indescribable, and none but a naturalist can understand the intense excitement I experienced when I at length captured it. On taking it out of my net and opening the glorious wings, my heart began to beat violently, the blood rushed to my head, and I felt much more like fainting than I have done when in apprehension of immediate death. I had a head-ache the rest of the day, so great was the excitement produced by what will appear to most people a very inadequate cause.

I had decided to return to Ternate in a week or two more, but this grand capture determined me to stay on till

I obtained a good series of the new butterfly, which I have since named Ornithoptera crœsus. The Mussænda bush was an admirable place, which I could visit every day on my way to the forest; and as it was situated in a dense thicket of shrubs and creepers, I set my man Lahi to clear a space all round it, so that I could easily get at any insect that might visit it. Afterwards, finding that it was often necessary to wait some time there, I had a little seat put up under a tree by the side of it, where I came every day to eat my lunch, and thus had half an hour's watching about noon, besides a chance as I passed it in the morning. In this way I obtained on an average one specimen a day for a long time, but more than half of these were females, and more than half the remainder worn or broken specimens, so that I should not have obtained many perfect males had I not found another station for them.

As soon as I had seen them come to flowers, I sent my man Lahi with a net on purpose to search for them, as they had also been seen at some flowering trees on the beach, and I promised him half a day's wages extra for every good specimen he could catch. After a day or two he brought me two very fair specimens, and told me he had caught them in the bed of a large rocky stream that descends from the mountains to the sea about a mile below the village. They flew down this river, settling occasionally on stones and rocks in the water, and he was obliged to wade up it or jump from rock to rock to get at them. I went with him one day, but found that the stream was far too rapid and the stones too slippery for me to do anything, so I left it entirely to him, and all the rest of the time we stayed in Batchian he used to be out all day, generally bringing me one, and on good days two or three specimens. I was thus able to bring away with me more than a hundred of both sexes, including perhaps twenty very fine males, though not more than five or six that were absolutely perfect.

My daily walk now led me, first about half a mile along the sandy beach, then through a sago swamp over a causeway of very shaky poles to the village of the Tomōré people. Beyond this was the forest with patches of new clearing, shady paths, and a considerable quantity of felled timber. I found this a very fair collecting ground, especially for beetles. The fallen trunks in the clearings abounded with golden Buprestidæ and curious Brenthidæ and longicorns, while in the forest I found abundance of the smaller Curculionidæ, many longicorns, and some fine green Carabidæ.

Butterflies were not abundant, but I obtained a few more of the fine blue Papilio, and a number of beautiful little Lycænidæ, as well as a single specimen of the very rare Papilio Wallacei, of which I had taken the hitherto unique specimen in the Aru Islands.

The most interesting birds I obtained here, were the beautiful blue kingfisher, Todiramphus diops; the fine green and purple doves, Ptilonopus superbus and P. iogaster, and several new birds of small size. My shooters still brought me in specimens of the Semioptera Wallacei, and I was greatly excited by the positive statements of several of the native hunters that another species of this bird existed, much handsomer and more remarkable. They declared that the plumage was glossy black, with metallic green breast as in my species, but that the white shoulder plumes were twice as long, and hung down far below the body of the bird. They declared that when hunting pigs or deer far in the forest they occasionally saw this bird, but that it was rare. I immediately offered twelve guilders (a pound) for a specimen; but all in vain, and I am to this day uncertain whether such a bird exists. Since I left, the German naturalist, Dr. Bernstein, stayed many months in the island with a large staff of hunters collecting for the Leyden Museum; and as he was not more successful than myself, we must consider either that the bird is very rare, or is altogether a myth.

Batchian is remarkable as being the most eastern point on the globe inhabited by any of the Quadrumana. A large black baboon-monkey (Cynopithecus nigrescens) is abundant in some parts of the forest. This animal has bare red callosities, and a rudimentary tail about an inch long—a mere fleshy tubercle, which may be very easily overlooked. It is the same species that is found all over the forests of Celebes, and as none of the other Mammalia of that island extend into Batchian I am inclined to suppose that this species has been accidentally introduced by the roaming Malays, who often carry about with them tame monkeys and other animals. This is rendered more probable by the fact that the animal is not found in Gilolo, which is only separated from Batchian by a very narrow strait. The introduction may have been very recent, as in a fertile and unoccupied island such an animal would multiply rapidly. The only other mammals obtained were an Eastern opossum, which Dr. Gray has described as *Cuscus ornatus*; the little flying opossum, *Belideus ariel*; a Civet cat, *Viverra zebetha*; and nine species of bats, most of the smaller ones being caught in the dusk with my butterfly net as they flew about before the house.

After much delay, owing to bad weather and the illness of one of my men, I determined to visit Kasserota (formerly the chief village), situated up a small stream, on an island close to the north coast of Batchian ; where I was told that many rare birds were found. After my boat was loaded and everything ready, three days of heavy squalls prevented our starting, and it was not till the 21st of March that we got away. Early next morning

we entered the little river, and in about an hour we
reached the Sultan's house, which I had obtained permission to use.   I was situated on the bank of the river,
and surrounded by a forest of fruit trees, among which
were some of the very loftiest and most graceful cocoa-nut
palms I have ever seen.   It rained nearly all that day,
and I could do little but unload and unpack.   Towards
the afternoon it cleared up, and I attempted to explore in
various directions, but found to my disgust that the only
path was a perfect mud swamp, along which it was almost
impossible to walk, and the surrounding forest so damp
and dark as to promise little in the way of insects.   I
found too on inquiry that the people here made no clearings, living entirely on sago, fruit, fish, and game ; and the
path only led to a steep rocky mountain equally imprac-
ticable and unproductive.   The next day I sent my men
to this hill, hoping it might produce some good birds ; but
they returned with only two common species, and I myself
had been able to get nothing, every little track I had
attempted to follow leading to a dense sago swamp.   I
saw that I should waste time by staying here, and determined to leave the following day.

This is one of those spots so hard for the European
naturalist to conceive, where with all the riches of a
tropical vegetation, and partly perhaps from the very
luxuriance of that vegetation, insects are as scarce as in

the most barren parts of Europe, and hardly more conspicuous. In temperate climates there is a tolerable
uniformity in the distribution of insects over those parts
of a country in which there is a similarity in the vegetation, any deficiency being easily accounted for by the
absence of wood or uniformity of surface. The traveller
hastily passing through such a country can at once pick
out a collecting ground which will afford him a fair
notion of its entomology. Here the case is different.
There are certain requisites of a good collecting ground
which can only be ascertained to exist by some days'
search in the vicinity of each village. In some places
there is no virgin forest, as at Djilolo and Sahoe; in
others there are no open pathways or clearings, as here.
At Batchian there are only two tolerable collecting places,
—the road to the coal mines, and the new clearings made
by the Tomōré people, the latter being by far the most
productive. I believe the fact to be that insects are pretty
uniformly distributed over these countries (where the
forests have not been cleared away), and are so scarce in
any one spot that searching for them is almost useless.
If the forest is all cleared away, almost all the insects
disappear with it ; but when small clearings and paths are
made, the fallen trees in various stages of drying and
decay, the rotting leaves, the loosening bark and the fun-
goid growths upon it, together with the flowers that appear

in much greater abundance where the light is admitted, are so many attractions to the insects for miles around, and cause a wonderful accumulation of species and individuals. When the entomologist can discover such a spot, he does more in a month than he could possibly do by a year's search in the depths of the undisturbed forest.

The next morning we left early, and reached the mouth of the little river in about an hour. It flows through a perfectly flat alluvial plain, but there are hills which approach it near the mouth. Towards the lower part, in a swamp where the salt-water must enter at high tides, were a number of elegant tree-ferns from eight to fifteen feet high. These are generally considered to be mountain plants, and rarely to occur on the equator at an elevation of less than one or two thousand feet. In Borneo, in the Aru Islands, and on the banks of the Amazon, I have observed them at the level of the sea, and think it probable that the altitude supposed to be requisite for them may have been deduced from facts observed in countries where the plains and lowlands are largely cultivated, and most of the indigenous vegetation destroyed. Such is the case in most parts of Java, India, Jamaica, and Brazil, where the vegetation of the tropics has been most fully explored.

Coming out to sea we turned northwards, and in about two hours' sail reached a few huts, called Langundi, where some Galela men had established themselves as collectors of gum-dammar, with which they made torches for the supply of the Ternate market. About a hundred yards back rises a rather steep hill, and a short walk having shown me that there was a tolerable path up it, I determined to stay here for a few days. Opposite us, and all along this coast of Batchian, stretches a row of fine islands completely uninhabited. Whenever I asked the reason why no one goes to live in them, the answer always was, " For fear of the Magindano pirates." Every year these scourges of the Archipelago wander in one direction or another, making their rendezvous on some uninhabited island, and carrying devastation to all the small settlements around ; robbing, destroying, killing, or taking captive all they mee with. Their long well-manned praus escape from the pursuit of any sailing vessel by pulling away right in the wind's eye, and the warning smoke of a steamer generally enables them to hide in some shallow bay, or narrow river, or forest-covered inlet, till the danger is passed. The only effectual way to put a stop to their depredations would be to attack them in their strongholds and villages, and compel them to give up piracy, and submit to strict surveillance. Sir James Brooke did this with the pirates of the north-west coast of Borneo, and deserves the thanks of the whole population of the Archipelago for having rid them of half their enemies.

All along the beach here, and in the adjacent strip of sandy lowland, is a remarkable display of Pandanaceæ or Screw-pines. Some are like huge branching candelabra, forty or fifty feet high, and bearing at the end of each branch a tuft of immense sword-shaped leaves, six or eight inches wide, and as many feet long. Others have a single unbranched stem, six or seven feet high, the upper part clothed with the spirally arranged leaves, and bearing a single terminal fruit as large as a swan's egg. Others of intermediate size have irregular clusters of rough red fruits, and all have more or less spiny-edged leaves and ringed stems. The young plants of the larger species have smooth glossy thick leaves, sometimes ten feet long and eight inches wide, which are used all over the Moluccas and New Guinea, to make "cocoyas" or sleeping mats, which are often very prettily ornamented with coloured patterns. Higher up on the hill is a forest of immense trees, among which those producing the resin called dammar (Dammara sp.) are abundant. The inhabitants of several small villages in Batchian are entirely engaged in searching for this product, and making it into torches by pounding it and filling it into tubes of palm leaves about a yard long, which are the only lights used by many of the natives. Sometimes the dammar accumulates in large masses of ten or twenty pounds weight, either attached to the trunk, or found buried in the

ground at the foot of the trees. The most extraordinary
trees of the forest are, however, a kind of fig, the aërial
roots of which form a pyramid near a hundred feet high,
terminating just where the tree branches out above, so that
there is no real trunk. This pyramid or cone is formed of
roots of every size, mostly descending in straight lines, but
more or less obliquely—and so crossing each other, and
connected by cross branches, which grow from one to
another; as to form a dense and complicated network, to
which nothing but a photograph could do justice (see illustration at Vol. I. page 130). The Kanary is also abundant in this forest, the nut of which has a very agreeable
flavour, and produces an excellent oil. The fleshy outer
covering of the nut is the favourite food of the great green
pigeons of these islands (Carpophaga perspicillata), and
their hoarse cooings and heavy flutterings among the
branches can be almost continually heard.

After ten days at Langundi, finding it impossible to get
the bird I was particularly in search of (the Nicobar
pigeon, or a new species allied to it), and finding no new
birds, and very few insects, I left early on the morning of
April 1st, and in the evening entered a river on the main
island of Batchian (Langundi, like Kasserota, being on a
distinct island), where some Malays and Galela men have a
small village, and have made extensive rice-fields and plantain grounds. Here we found a good house near the river

bank, where the water was fresh and clear, and the owner, a respectable Batchian Malay, offered me sleeping room and the use of the verandah if I liked to stay. Seeing forest all round within a short distance, I accepted his offer, and the next morning before breakfast walked out to explore, and on the skirts of the forest captured a few interesting insects.

Afterwards, I found a path which led for a mile or more through a very fine forest, richer in palms than any I had seen in the Moluccas. One of these especially attracted my attention from its elegance. The stem was not thicker than my wrist, yet it was very lofty, and bore clusters of bright red fruit. It was apparently a species of Areca. Another of immense height closely resembled in appearance the Euterpes of South America. Here also grew the fan-leafed palm, whose small, nearly entire leaves are used to make the dammar torches, and to form the water-buckets in universal use. During this walk I saw near a dozen species of palms, as well as two or three Pandani different from those of Langundi. There were also some very fine climbing ferns and true wild Plantains (Musa), bearing an edible fruit not so large as one's thumb, and consisting of a mass of seeds just covered with pulp and skin. The people assured me they had tried the experiment of sowing and cultivating this species, but could not improve it. They probably did not

grow it in sufficient quantity, and did not persevere sufficiently long.

Batchian is an island that would perhaps repay the
researches of a botanist better than any other in the
whole Archipelago. It contains a great variety of surface and of soil, abundance of large and small streams,
many of which are navigable for some distance, and there
being no savage inhabitants, every part of it can be visited
with perfect safety. It possesses gold, copper, and coal,
hot springs and geysers, sedimentary and volcanic rocks
and coralline limestone, alluvial plains, abrupt hills and
lofty mountains, a moist climate, and a grand and luxuriant
forest vegetation.

The few days I stayed here produced me several new
insects, but scarcely any birds. Butterflies and birds are
in fact remarkably scarce in these forests. One may walk
a whole day and not see more than two or three species of
either. In everything but beetles, these eastern islands
are very deficient compared with the western (Java,
Borneo, &c.), and much more so if compared with the
forests of South America, where twenty or thirty species
of butterflies may be caught every day, and on very good
days a hundred, a number we can hardly reach here in
months of unremitting search. In birds there is the same
difference. In most parts of tropical America we may
always find some species of woodpecker tanager, bush-

shrike, chatterer, trogon, toucan, cuckoo, and tyrant-fly-catcher; and a few days' active search will produce more variety than can be here met with in as many months. Yet, along with this poverty of individuals and of species, there are in almost every class and order, some one or two species of such extreme beauty or singularity, as to vie with, or even surpass, anything that even South America can produce.

One afternoon when I was arranging my insects, and surrounded by a crowd of wondering spectators, I showed one of them how to look at a small insect with a hand-lens, which caused such evident wonder that all the rest wanted to see it too. I therefore fixed the glass firmly to a piece of soft wood at the proper focus, and put under it a little spiny beetle of the genus Hispa, and then passed it round for examination. The excitement was immense. Some declared it was a yard long; others were frightened, and instantly dropped it, and all were as much astonished, and made as much shouting and gesticulation, as children at a pantomime, or at a Christmas exhibition of the oxy-hydrogen microscope. And all this excitement was produced by a little pocket lens, an inch and a half focus, and therefore magnifying only four or five times, but which to their unaccustomed eyes appeared to enlarge a hundred fold.

On the last day of my stay here, one of my hunters

succeeded in finding and shooting the beautiful Nicobar
pigeon, of which I had been so long in search. None
of the residents had ever seen it, which shows that it is
rare and shy. My specimen was a female in beautiful
condition, and the glossy coppery and green of its plumage,
the snow-white tail and beautiful pendent feathers of the
neck, were greatly admired. I subsequently obtained a
specimen in New Guinea, and once saw it in the Kaióa
islands. It is found also in some small islands near
Macassar, in others near Borneo, and in the Nicobar
islands, whence it receives its name. It is a ground
feeder, only going upon trees to roost, and is a very
heavy fleshy bird. This may account for the fact of its
being found chiefly on very small islands, while in the
western half of the Archipelago, it seems entirely absent
from the larger ones. Being a ground feeder it is subject
to the attacks of carnivorous quadrupeds, which are not
found in the very small islands. Its wide distribution over
the whole length of the Archipelago, from extreme west to
east, is however very extraordinary, since, with the exception of a few of the birds of prey, not a single land bird
has so wide a range. Ground-feeding birds are generally
deficient in power of extended flight, and this species is so
bulky and heavy that it appears at first sight quite unable
to fly a mile. A closer examination shows, however, that
its wings are remarkably large, perhaps in proportion to

its size larger than those of any other pigeon, and its
pectoral muscles are immense. A fact communicated to
me by the son of my friend Mr. Duivenboden of Ternate,
would show that, in accordance with these peculiarities of
structure, it possesses the power of flying long distances.
Mr. D. established an oil factory on a small coral island, a
hundred miles north of New Guinea, with no intervening
land. After the island had been settled a year, and
traversed in every direction, his son paid it a visit ;
and just as the schooner was coming to an anchor, a bird
was seen flying from seaward which fell into the water
exhausted before it could reach the shore. A boat was
sent to pick it up, and it was found to be a Nicobar
pigeon, which must have come from New Guinea, and
flown a hundred miles, since no such bird previously
inhabited the island.

This is certainly a very curious case of adaptation to
an unusual and exceptional necessity. The bird does
not ordinarily require great powers of flight, since it
lives in the forest, feeds on fallen fruits, and roosts
in low trees like other ground pigeons. The majority
of the individuals, therefore, can never make full use
of their enormously powerful wings, till the exceptional
case occurs of an individual being blown out to sea,
or driven to emigrate by the incursion of some carnivorous animal, or the pressure of scarcity of food. A

modification exactly opposite to that which produced the wingless birds (the Apteryx, Cassowary, and Dodo), appears to have here taken place; and it is curious that in both cases an insular habitat should have been the moving cause. The explanation is probably the same as that applied by Mr. Darwin to the case of the Madeira beetles, many of which are wingless, while some of the winged ones have the wings better developed than the same species on the continent. It was advantageous to these insects either never to fly at all, and thus not run the risk of being blown out to sea, or to fly so well as to be able either to return to land, or to migrate safely to the continent. Bad flying was worse than not flying at all. So, while in such islands as New Zealand and Mauritius, far from all land, it was safer for a ground-feeding bird not to fly at all, and the short-winged individuals continually surviving, prepared the way for a wingless group of birds; in a vast Archipelago thickly strewn with islands and islets it was advantageous to be able occasionally to migrate, and thus the long and strong-winged varieties maintained their existence longest, and ultimately supplanted all others, and spread the race over the whole Archipelago.

Besides this pigeon, the only new bird I obtained during the trip was a rare goat-sucker (Batrachostomus crinifrons), the only species of the genus yet found in the Moluccas.

Among my insects the best were the rare Pieris aruna, of
a rich chrome yellow colour, with a black border and
remarkable white antennæ—perhaps the very finest butterfly of the genus; and a large black wasp-like insect,
with immense jaws like a stag-beetle, which has been
named Megachile pluto by Mr. F. Smith. I collected
about a hundred species of beetles quite new to me, but
mostly very minute, and also many rare and handsome
ones which I had already found in Batchian. On the
whole I was tolerably satisfied with my seventeen days'
excursion, which was a very agreeable one, and enabled
me to see a good deal of the island. I had hired a roomy
boat, and brought with me a small table and my rattan
chair. These were great comforts, as, wherever there was
a roof, I could immediately instal myself, and work and
eat at ease. When I could not find accommodation on
shore I slept in the boat, which was always drawn up
on the beach if we stayed for a few days at one spot.

On my return to Batchian I packed up my collections,
and prepared for my return to Ternate. When I first
came I had sent back my boat by the pilot, with two or
three other men who had been glad of the opportunity.
I now took advantage of a Government boat which had
just arrived with rice for the troops, and obtained permission to return in her, and accordingly started on the
13th of April, having resided only a week short of six

months on the island of Batchian. The boat was one of
the kind called "Kora-kora," quite open, very low, and
about four tons burthen. It had outriggers of bamboo
about five feet off each side, which supported a bamboo
platform extending the whole length of the vessel. On
the extreme outside of this sit the twenty rowers, while
within was a convenient passage fore and aft. The middle
portion of the boat was covered with a thatch-house, in
which baggage and passengers are stowed; the gunwale
was not more than a foot above water, and from the great
top and side weight, and general clumsiness, these boats
are dangerous in heavy weather, and are not unfrequently
lost. A triangle mast and mat sail carried us on when
the wind was favourable, which (as usual) it never was,
although, according to the monsoon, it ought to have been.
Our water, carried in bamboos, would only last two days,
and as the voyage occupied seven, we had to touch at
a great many places. The captain was not very energetic,
and the men rowed as little as they pleased, or we might
have reached Ternate in three days, having had fine
weather and little wind all the way.

There were several passengers besides myself : three
or four Javanese soldiers, two convicts whose time had
expired (one, curiously enough, being the man who had
stolen my cash-box and keys), the schoolmaster's wife
and a servant going on a visit to Ternate, and a Chinese

trader going to buy goods. We had to sleep all together in the cabin, packed pretty close; but they very civilly allowed me plenty of room for my mattrass, and we got on very well together. There was a little cook-house in the bows, where we could boil our rice and make our coffee, every one of course bringing his own provisions, and arranging his meal-times as he found most convenient. The passage would have been agreeable enough but for the dreadful "tom-toms," or wooden drums, which are beaten incessantly while the men are rowing. Two men were engaged constantly at them, making a fearful din the whole voyage. The rowers are men sent by the Sultan of Ternate. They get about three-pence a day, and find their own provisions. Each man had a strong wooden "betel" box, on which he generally sat, a sleeping-mat, and a change of clothes—rowing naked, with only a sarong or a waist-cloth. They sleep in their places, covered with their mat, which keeps out the rain pretty well. They chew betel or smoke cigarettes incessantly; eat dry sago and a little salt fish; seldom sing while rowing, except when excited and wanting to reach a stopping-place, and do not talk a great deal. They are mostly Malays, with a sprinkling of Alfuros from Gilolo, and Papuans from Guebe or Waigiou.

One afternoon we stayed at Makian; many of the men went on shore, and a great deal of plantains, bananas, and

other fruits were brought on board. We then went on a
little way, and in the evening anchored again. When
going to bed for the night, I put out my candle, there
being still a glimmering lamp burning, and, missing my
handkerchief, thought I saw it on a box which formed one
side of my bed, and put out my hand to take it. I quickly
drew back on feeling something cool and very smooth,
which moved as I touched it. "Bring the light, quick," I
cried ; "here's a snake." And there he was, sure enough,
nicely coiled up, with his head just raised to inquire who
had disturbed him. It was now necessary to catch or kill
him neatly, or he would escape among the piles of miscellaneous luggage, and we should hardly sleep comfortably
One of the ex-convicts volunteered to catch him with his
hand wrapped up in a cloth, but from the way he went
about it I saw he was nervous and would let the thing go,
so I would not allow him to make the attempt. I then got
a chopping-knife, and carefully moving my insect nets,
which hung just over the snake and prevented me getting
a free blow, I cut him quietly across the back, holding
him down while my boy with another knife crushed his
head. On examination, I found he had large poison
fangs, and it is a wonder he did not bite me when I
first touched him.

Thinking it very unlikely that two snakes had got on
board at the same time, I turned in and went to sleep ;

but having all the time a vague dreamy idea that I might put my hand on another one, I lay wonderfully still, not turning over once all night, quite the reverse of my usual habits. The next day we reached Ternate, and I ensconced myself in my comfortable house, to examine all my treasures, and pack them securely for the voyage home.

CHAPTER XXV.

CERAM, GORAM, AND THE MATABELLO ISLANDS.

(OCTOBER 1859 TO JUNE 1860.)

I LEFT Amboyna for my first visit to Ceram at three o'clock in the morning of October 29th, after having been delayed several days by the boat's crew, who could not be got together. Captain Van der Beck, who gave me a passage in his boat, had been running after them all day, and at midnight we had to search for two of my men who had disappeared at the last moment. One we found at supper in his own house, and rather tipsy with his parting libations of arrack, but the other was gone across the bay, and we were obliged to leave without him. We stayed some hours at two villages near the east end of Amboyna, at one of which we had to discharge some wood for the missionaries' house, and on the third afternoon reached Captain Van der Beck's plantation, situated at Hatosúa, in that part of Ceram opposite to the island of Amboyna. This was a clearing in flat and rather swampy forest, about twenty acres in extent, and mostly planted with cacao and

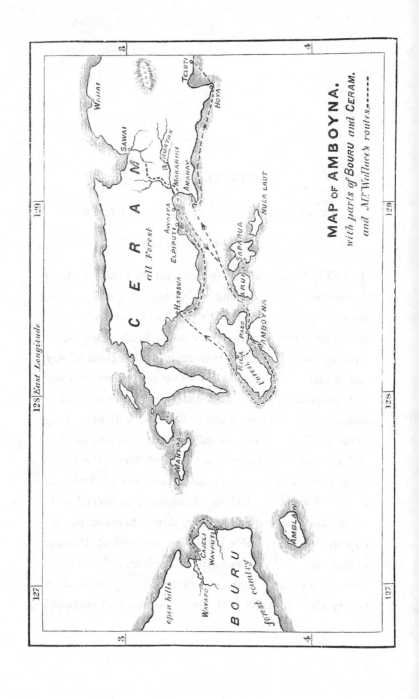

MAP OF **AMBOYNA.**

with parts of **Bouru** *and* **Ceram.**

and Mr Wallace's routes. ------

tobacco. Besides a small cottage occupied by the workmen, there was a large shed for tobacco drying, a corner of which was offered me; and thinking from the look of the place that I should find good collecting ground here, I fitted up temporary tables, benches, and beds, and made all preparations for some weeks' stay. A few days, however, served to show that I should be disappointed. Beetles were tolerably abundant, and I obtained plenty of fine long-horned Anthribidæ and pretty Longicorns, but they were mostly the same species as I had found during my first short visit to Amboyna. There were very few paths in the forest, which seemed poor in birds and butterflies, and day after day my men brought me nothing worth notice. I was therefore soon obliged to think about changing my locality, as I could evidently obtain no proper notion of the productions of the almost entirely unexplored island of Ceram by staying in this place.

I rather regretted leaving, because my host was one of the most remarkable men and most entertaining companions I had ever met with. He was a Fleming by birth, and, like so many of his countrymen, had a wonderful talent for languages. When quite a youth he had accompanied a Government official who was sent to report on the trade and commerce of the Mediterranean, and had acquired the colloquial language of every place they stayed a few weeks at. He had afterwards made voyages

to St. Petersburg, and to other parts of Europe, including
a few weeks in London, and had then come out to the
East, where he had been for some years trading and
speculating in the various islands. He now spoke Dutch,
French, Malay, and Javanese, all equally well; English
with a very slight accent, but with perfect fluency, and a
most complete knowledge of idiom, in which I often tried
to puzzle him in vain. German and Italian were also
quite familiar to him, and his acquaintance with European
languages included Modern Greek, Turkish, Russian, and
colloquial Hebrew and Latin. As a test of his power, I
may mention that he had made a voyage to the out-of-the-
way island of Salibaboo, and had stayed there trading a
few weeks. As I was collecting vocabularies, he told me
he thought he could remember some words, and dictated
a considerable number. Some time after I met with a
short list of words taken down in those islands, and in
every case they agreed with those he had given me. He
used to sing a Hebrew drinking-song, which he had learned
from some Jews with whom he had once travelled, and
astonished by joining in their conversation, and had a
never-ending fund of tale and anecdote about the people
he had met and the places he had visited.

In most of the villages of this part of Ceram are schools
and native schoolmasters, and the inhabitants have been
long converted to Christianity. In the larger villages

there are European missionaries; but there is little or no external difference between the Christian and Alfuro villages, nor, as far as I have seen, in their inhabitants. The people seem more decidedly Papuan than those of Gilolo. They are darker in colour, and a number of them have the frizzly Papuan hair; their features also are harsh and prominent, and the women in particular are far less engaging than those of the Malay race. Captain Van der Beck was never tired of abusing the inhabitants of these Christian villages as thieves, liars, and drunkards, besides being incorrigibly lazy. In the city of Amboyna my friends Doctors Mohnike and Doleschall, as well as most of the European residents and traders, made exactly the same complaint, and would rather have Mahometans for servants, even if convicts, than any of the native Christians. One great cause of this is the fact, that with the Mahometans temperance is a part of their religion, and has become so much a habit that practically the rule is never transgressed. One fertile source of want, and one great incentive to idleness and crime, is thus present with the one class, but absent in the other; but besides this the Christians look upon themselves as nearly the equals of the Europeans, who profess the same religion, and as far superior to the followers of Islam, and are therefore prone to despise work, and to endeavour to live by trade, or by cultivating their own land. It need hardly be said that with people in this low state of civilization religion is almost wholly ceremonial, and that neither are the doctrines of Christianity comprehended, nor its moral precepts obeyed. At the same time, as far as my own experience goes, I have found the better class of " Orang Sirani" as civil, obliging, and industrious as the Malays, and only inferior to them from their tendency to get intoxicated.

Having written to the Assistant Resident of Saparua (who has jurisdiction over the opposite part of the coast of Ceram) for a boat to pursue my journey, I received one rather larger than necessary with a crew of twenty men. I therefore bade adieu to my kind friend Captain Van der Beck, and left on the evening after its arrival for the village of Elpiputi, which we reached in two days. I had intended to stay here, but not liking the appearance of the place, which seemed to have no virgin forest near it, I determined to proceed about twelve miles further up the bay of Amahay, to a village recently formed, and inhabited by indigenes from the interior, and where some extensive cacao plantations were being made by some gentlemen of Amboyna. I reached the place (called Awaiya) the same afternoon, and with the assistance of Mr. Peters (the manager of the plantations) and the native chief, obtained a small house, got all my things on shore, and paid and discharged my twenty boatmen, two of whom had almost driven me to distraction by beating tom-toms the whole voyage.

I found the people here very nearly in a state of nature, and going almost naked. The men wear their frizzly hair gathered into a flat circular knot over the left temple, which has a very knowing look, and in their ears cylinders of wood as thick as one's finger, and coloured red at the ends. Armlets and anklets of woven grass or of silver, with necklaces of beads or of small fruits, complete their attire. The women wear similar ornaments, but have their hair loose. All are tall, with a dark brown skin, and well marked Papuan physiognomy. There is an Amboyna schoolmaster in the village, and a good number of children attend school every morning. Such of the inhabitants as have become Christians may be known by their wearing their hair loose, and adopting to some extent the native Christian dress—trousers and a loose shirt. Very few speak Malay, all these coast villages having been recently formed by inducing natives to leave the inaccessible interior. In all the central part of Ceram there now remains only one populous village in the mountains. Towards the east and the extreme west are a few others, with which exceptions all the inhabitants of Ceram are collected on the coast. In the northern and eastern districts they are mostly Mahometans, while on the southwest coast, nearest Amboyna, they are nominal Christians.

In all this part of the Archipelago the Dutch make very praiseworthy efforts to improve the condition of the aborigines by establishing schoolmasters in every village (who are mostly natives of Amboyna or Saparua, who have been instructed by the resident missionaries), and by employing native vaccinators to prevent the ravages of small-pox. They also encourage the settlement of Europeans, and the formation of new plantations of cacao and coffee, one of the best means of raising the condition of the natives, who thus obtain work at fair wages, and have the opportunity of acquiring something of European tastes and habits.

My collections here did not progress much better than at my former station, except that butterflies were a little more plentiful, and some very fine species were to be found in the morning on the sea-beach, sitting so quietly on the wet sand that they could be caught with the fingers. In this way I had many fine specimens of Papilios brought me by the children. Beetles, however, were scarce, and birds still more so, and I began to think that the hand-some species which I had so often heard were found in Ceram must be entirely confined to the eastern extremity of the island.

A few miles further north, at the head of the Bay of Amahay, is situated the village of Makariki, from whence there is a native path quite across the island to the north coast. My friend Mr. Rosenberg, whose acquaintance I had made at New Guinea, and who was now the Government superintendent of all this part of Ceram, returned from Wahai, on the north coast, after I had been three weeks at Awaiya, and showed me some fine butterflies he had obtained on the mountain streams in the interior. He indicated a spot about the centre of the island where he thought I might advantageously stay a few days.   I accordingly visited Makariki with him the next day, and he instructed the chief of the village to furnish me with men to carry my baggage, and accompany me on my excursion. As the people of the village wanted to be at home on Christmas-day, it was necessary to start as soon as possible; so we agreed that the men should be ready in two days, and I returned to make my arrangements.

I put up the smallest quantity of baggage possible for a six days' trip, and on the morning of December 18th we left Makariki, with six men carrying my baggage and their own provisions, and a lad from Awaiya, who was accustomed to catch butterflies for me. My two Amboyna hunters I left behind to shoot and skin what birds they could while I was away. Quitting the village, we first walked briskly for an hour through a dense tangled undergrowth, dripping wet from a storm of the previous night, and full of mud holes. After crossing several small streams we reached one of the largest rivers in

Ceram, called Ruatan, which it was necessary to cross. It was both deep and rapid. The baggage was first taken over, parcel by parcel, on the men's heads, the water reaching nearly up to their armpits, and then two men returned to assist me. The water was above my waist, and so strong that I should certainly have been carried off my feet had I attempted to cross alone; and it was a matter of astonishment to me how the men could give me any assistance, since I found the greatest difficulty in getting my foot down again when I had once moved it off the bottom. The greater strength and grasping power of their feet, from going always barefoot, no doubt gave them a surer footing in the rapid water.

After well wringing out our wet clothes and putting them on, we again proceeded along a similar narrow forest track as before, choked with rotten leaves and dead trees, and in the more open parts overgrown with tangled vegetation. Another hour brought us to a smaller stream flowing in a wide gravelly bed, up which our road lay. Here we stayed half an hour to breakfast, and then went on, continually crossing the stream, or walking on its stony and gravelly banks, till about noon, when it became rocky and enclosed by low hills. A little further we entered a regular mountain-gorge, and had to clamber over rocks, and every moment cross and recross the water, or take short cuts through the forest. This was fatiguing work ; and about three in the afternoon, the sky being overcast, and thunder in the mountains indicating an approaching storm, we had to look out for a camping place, and soon after reached one of Mr. Rosenberg's old ones. The skeleton of his little sleeping-hut remained, and my men cut leaves and made a hasty roof just as the rain commenced. The baggage was covered over with leaves, and the men sheltered themselves as they could till the storm was over, by which time a flood came down the river, which effectually stopped our further march, even had we wished to proceed. We then lighted fires ; I made some coffee, and my men roasted their fish and plantains, and as soon as it was dark, we made ourselves comfortable for the night.

Starting at six the next morning, we had three hours of the same kind of walking, during which we crossed the river at least thirty or forty times, the water being generally knee-deep. This brought us to a place where the road left the stream, and here we stopped to breakfast. We then had a long walk over the mountain, by a tolerable path, which reached an elevation of about fifteen hundred feet above the sea. Here I noticed one of the smallest and most elegant tree ferns I had ever seen, the stem being scarcely thicker than my thumb, yet reaching a height of fifteen or twenty feet. I also caught a new butterfly of the genus Pieris, and a magnificent female specimen of Papilio gambrisius, of which I had hitherto only found the males, which are smaller and very different in colour. Descending the other side of the ridge, by a very steep path, we reached another river at a spot which is about the centre of the island, and which was to be our resting-place for two or three days. In a couple of hours my men had built a little sleeping-shed for me, about eight feet by four, with a bench of split poles, they themselves occupying two or three smaller ones, which had been put up by former passengers.

The river here was about twenty yards wide, running over a pebbly and sometimes a rocky bed, and bordered by steep hills with occasionally flat swampy spots between their base and the stream. The whole country was one dense, unbroken, and very damp and gloomy virgin forest. Just at our resting-place there was a little bush-covered island in the middle of the channel, so that the opening in the forest made by the river was wider than usual, and allowed a few gleams of sunshine to penetrate. Here there were several handsome butterflies flying about, the finest of which, however, escaped me, and I never saw it again during my stay. In the two days and a half which we remained here, I wandered almost all day up and down the stream, searching after butterflies, of which I got, in all, fifty or sixty specimens, with several species quite new to me. There were

many others which I saw only once, and did not capture,
causing me to regret that there was no village in these
interior valleys where I could stay a month. In the early
part of each morning I went out with my gun in search of
birds, and two of my men were out almost all day after
deer; but we were all equally unsuccessful, getting absolutely nothing the whole time we were in the forest.
The only good bird seen was the fine Amboyna lory, but
these were always too high to shoot; besides this, the
great Moluccan hornbill, which I did not want, was
almost the only bird met with. I saw not a single ground-
thrush, or kingfisher, or pigeon; and, in fact, have never
been in a forest so utterly desert of animal life as this
appeared to be. Even in all other groups of insects,
except butterflies, there was the same poverty. I had
hoped to find some rare tiger beetles, as I had done in
similar situations in Celebes; but, though I searched
closely in forest, river-bed, and mountain-brook, I could
find nothing but the two common Amboyna species.
Other beetles there were absolutely none.

The constant walking in water, and over rocks and
pebbles, quite destroyed the two pair of shoes I brought
with me, so that, on my return, they actually fell to
pieces, and the last day I had to walk in my stockings
very painfully, and reached home quite lame. On our
way back from Makariki, as on our way there, we had

storm and rain at sea, and we arrived at Awaiya late in the evening, with all our baggage drenched, and ourselves thoroughly uncomfortable. All the time I had been in Ceram I had suffered much from the irritating bites of an invisible acarus, which is worse than mosquitoes, ants, and every other pest, because it is impossible to guard against them. This last journey in the forest left me covered from head to foot with inflamed lumps, which, after my return to Amboyna, produced a serious disease, confining me to the house for nearly two months,—a not very pleasant memento of my first visit to Ceram, which terminated with the year 1859.

It was not till the 24th of February, 1860, that I started again, intending to pass from village to village along the coast, staying where I found a suitable locality. I had a letter from the Governor of the Moluccas, requesting all the chiefs to supply me with boats and men to carry me on my journey. The first boat took me in two days to Amahay, on the opposite side of the bay to Awaiya. The chief here, wonderful to relate, did not make any excuses for delay, but immediately ordered out the boat which was to carry me on, put my baggage on board, set up mast and sails after dark, and had the men ready that night; so that we were actually on our way at five the next morning,—a display of energy and activity I scarcely ever saw before in a native chief on such an occasion. We

touched at Cepa, and stayed for the night at Tamilan, the
first two Mahometan villages on the south coast of Ceram.
The next day, about noon, we reached Hoya, which was as
far as my present boat and crew were going to take me.
The anchorage is about a mile east of the village, which is
faced by coral reefs, and we had to wait for the evening
tide to move up and unload the boat into the strange
rotten wooden pavilion kept for visitors.

There was no boat here large enough to take my
baggage; and although two would have done very well,
the Rajah insisted upon sending four. The reason of this I
found was, that there were four small villages under his
rule, and by sending a boat from each he would avoid the
difficult task of choosing two and letting off the others. I
was told that at the next village of Teluti there were
plenty of Alfuros, and that I could get abundance of lories
and other birds. The Rajah declared that black and yellow
lories and black cockatoos were found there; but I am inclined to think he knew very well he.was telling me lies,
and that it was only a scheme to satisfy me with his plan
of taking me to that village, instead of a day's journey
further on, as I desired. Here, as at most of the villages,
I was asked for spirits, the people being mere nominal
Mahometans, who confine their religion almost entirely to
a disgust at pork, and a few other forbidden articles of food.
The next morning, after much trouble, we got our cargoes

loaded, and had a delightful row across the deep bay of
Teluti, with a view of the grand central mountain-range of
Ceram. Our four boats were rowed by sixty men, with
flags flying and tom-toms beating, as well as very vigorous
shouting and singing to keep up their spirits. The sea was
smooth, the morning bright, and the whole scene very
exhilarating. On landing, the Orang-kaya and several of
the chief men, in gorgeous silk jackets, were waiting to
receive us, and conducted me to a house prepared for my
reception, where I determined to stay a few days, and see
if the country round produced anything new.

My first inquiries were about the lories, but I could get
very little satisfactory information. The only kinds known
were the ring-necked lory and the common red and green
lorikeet, both common at Amboyna. Black lories and
cockatoos were quite unknown. The Alfuros resided in the
mountains five or six days' journey away, and there were
only one or two live birds to be found in the village, and
these were worthless. My hunters could get nothing but
a few common birds ; and notwithstanding fine mountains,
luxuriant forests, and a locality a hundred miles eastward,
I could find no new insects, and extremely few even of the
common species of Amboyna and West Ceram. It was
evidently no use stopping at such a place, and I was
determined to move on as soon as possible.

The village of Teluti is populous, but straggling and very

dirty.   Sago trees here cover the mountain side, instead of
growing as usual in low swamps ; but a closer examination
shows that they grow in swampy patches, which have
formed among the loose rocks that cover the ground, and
which are kept constantly full of moisture by the rains, and
by the abundance of rills which trickle down among them.
This sago forms almost the whole subsistence of the inhabitants, who appear to cultivate nothing but a few small
patches of maize and sweet potatoes.   Hence, as before
explained, the scarcity of insects.   The Orang-kaya has
fine clothes, handsome lamps, and other expensive
European goods, yet lives every day on sago and fish as
miserably as the rest.

   After three days in this barren place I left on the morning of March 6th, in two boats of the same size as those
which had brought me to Teluti.   With some difficulty
I had obtained permission to take these boats on to Tobo,
where I intended to stay a while, and therefore got on
pretty quickly, changing men at the village of Laiemu,
and arriving in a heavy rain at Ahtiago.   As there was a
good deal of surf here, and likely to be more if the wind
blew hard during the night, our boats were pulled up on
the beach ; and after supping at the Orang-kaya's house, and
writing down a vocabulary of the language of the Alfuros,
who live in the mountains inland I returned to sleep in
the boat.   Next morning we proceeded, changing men at

Warenama, and again at Hatometen, at both of which places there was much surf and no harbour, so that the men had to go on shore and come on board by swimming. Arriving in the evening of March 7th at Batuassa, the first village belonging to the Rajah of Tobo, and under the government of Banda, the surf was very heavy, owing to a strong westward swell. We therefore rounded the rocky point on which the village was situated, but found it very little better on the other side. We were obliged, however, to go on shore here; and waiting till the people on the beach had made preparations, by placing a row of logs from the water's edge on which to pull up our boats, we rowed as quickly as we could straight on to them, after watching till the heaviest surfs had passed. The moment we touched ground our men all jumped out, and, assisted by those on shore, attempted to haul up the boat high and dry, but not having sufficient hands, the surf repeatedly broke into the stern. The steepness of the beach, however, prevented any damage being done, and the other boat having both crews to haul at it, was got up without difficulty.

The next morning, the water being low, the breakers were at some distance from shore, and we had to watch for a smooth moment after bringing the boats to the water's edge, and so got safely out to sea. At the two next villages, Tobo and Ossong, we also took in fresh men, who came swimming through the surf; and at the latter place the Rajah came on board and accompanied me to Kissa-laut, where he has a house which he lent me during my stay. Here again was a heavy surf, and it was with great difficulty we got the boats safely hauled up. At Amboyna I had been promised at this season a calm sea and the wind off shore, but in this case, as in every other, I had been unable to obtain any reliable information as to the winds and seasons of places distant two or three days' journey. It appears, however, that owing to the general direction of the island of Ceram (E.S.E. and W.N.W.), there is a heavy surf and scarcely any shelter on the south coast during the west monsoon, when alone a journey to the eastward can be safely made ; while during the east monsoon, when I proposed to return along the north coast to Wahai, I should probably find that equally exposed and dangerous. But although the general direction of the west monsoon in the Banda sea causes a heavy swell, with bad surf on the coast, yet we had little advantage of the wind ; for, owing I suppose to the numerous bays and headlands, we had contrary south-east or even due east winds all the way, and had to make almost the whole distance from Amboyna by force of rowing. We had therefore all the disadvantages, and none of the advantages, of this west monsoon, which I was told would insure me a quick and pleasant journey.

I was delayed at Kissa-laut just four weeks, although after the first three days I saw that it would be quite useless for me to stay, and begged the Rajah to give me a prau and men to carry me on to Goram. But instead of getting one close at hand, he insisted on sending several miles off; and when after many delays it at length arrived, it was altogether unsuitable and too small to carry my baggage. Another was then ordered to be brought immediately, and was promised in three days, but double that time elapsed and none appeared, and we were obliged at length to get one at the adjoining village, where it might have been so much more easily obtained at first. Then came caulking and covering over, and quarrels between the owner and the Rajah's men, which occupied more than another ten days, during all which time I was getting absolutely nothing, finding this part of Ceram a perfect desert in zoology, although a most beautiful country, and with a very luxuriant vegetation. It was a complete puzzle, which to this day I have not been able to understand; the only thing I obtained worth notice during my month's stay here being a few good land shells.

At length, on April 4th, we succeeded in getting away in our little boat of about four tons burthen, in which my numerous boxes were with difficulty packed so as to leave sleeping and cooking room. The craft could not boast an ounce of iron or a foot of rope in any part of its construction, nor a morsel of pitch or paint in its decoration. The planks were fastened together in the usual

ingenious way with pegs and rattans. The mast was a
bamboo triangle, requiring no shrouds, and carrying a long
mat sail; two rudders were hung on the quarters by rattans, the anchor was of wood, and a long and thick rattan
served as a cable. Our crew consisted of four men, whose
sole accommodation was about three feet by four in the
bows and stern, with the sloping thatch roof to stretch
themselves upon for a change. We had nearly a hundred
miles to go, fully exposed to the swell of the Banda sea,
which is sometimes very considerable; but we luckily had
it calm and smooth, so that we made the voyage in comparative comfort.

On the second day we passed the eastern extremity of
Ceram, formed of a group of hummocky limestone hills;
and, sailing by the islands of Kwammer and Keffing, both
thickly inhabited, came in sight of the little town of Kilwaru, which appears to rise out of the sea like a rustic
Venice. This place has really a most extraordinary appearance, as not a particle of land or vegetation can be
seen, but a long way out at sea a large village seems to
float upon the water. There is of course a small island of
several acres in extent; but the houses are built so closely
all round it upon piles in the water, that it is completely
hidden. It is a place of great traffic, being the emporium
for much of the produce of these Eastern seas, and is the
residence of many Bugis and Ceramese traders, and appears

to have been chosen on account of its being close to the
only deep channel between the extensive shoals of Ceram-
laut and those bordering the east end of Ceram. We now
had contrary east winds, and were obliged to pole over the
shallow coral reefs of Ceram-laut for nearly thirty miles.
The only danger of our voyage was just at its termination,
for as we were rowing towards Manowolko, the largest of
the Goram group, we were carried out so rapidly by a
strong westerly current, that I was almost certain at one
time we should pass clear of the island; in which case
our situation would have been both disagreeable and
dangerous, as, with the east wind which had just set in,
we might have been unable to return for many days, and
we had not a day's water on board. At the critical
moment I served out some strong spirits to my men, which
put fresh vigour into their arms, and carried us out of
the influence of the current before it was too late.

MANOWOLKO, GORAM GROUP.

On arriving at Manowolko, we found the Rajah was at
the opposite island of Goram; but he was immediately sent
for, and in the meantime a large shed was given for our
accommodation. At night the Rajah came, and the next
day I had a visit from him, and found, as I expected, that
I had already made his acquaintance three years before at

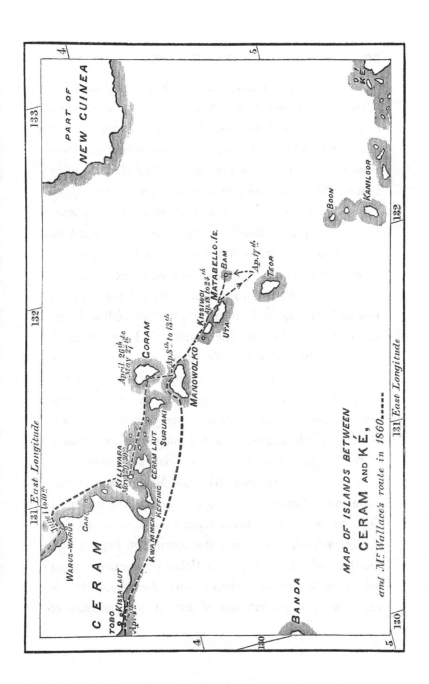

PART OF
NEW CUINEA

133

132 East Longitude

131 East Longitude

CERAM

WARUS-WARUS
CAH.
TOBO
KISSA LAUT
Ap.11th
KILLWARA
May 20.30th
June 10th
KWAMMER
KEFFING
CERAM LAUT
SURURAKI

CORAM
April 29th to
May 27th
Ap.8th to 13th

MANOWOLKO

KISSIWOI
Ap.18 to 24th
MATABELLO.Is.
UTA
BAM
Ap.17th
TEOR

BOON

KANHLOOR
K.

BANDA

130

131 East Longitude

132

MAP OF ISLANDS BETWEEN
CERAM AND KÉ,
and Mr Wallace's route in 1860.----

Aru. He was very friendly, and we had a long talk; but when I begged for a boat and men to take me on to Ké, he made a host of difficulties. There were no praus, as all had gone to Ké or Aru; and even if one were found, there were no men, as it was the season when all were away trading. But he promised to see about it, and I was obliged to wait. For the next two or three days there was more talking and more difficulties were raised, and I had time to make an examination of the island and the people.

Manowolko is about fifteen miles long, and is a mere upraised coral-reef. Two or three hundred yards inland rise cliffs of coral rock, in many parts perpendicular, and one or two hundred feet high; and this, I was informed, is characteristic of the whole island, in which there is no other kind of rock, and no stream of water. A few cracks and chasms furnish paths to the top of these cliffs, where there is an open undulating country, in which the chief vegetable grounds of the inhabitants are situated.

The people here—at least the chief men—were of a much purer Malay race than the Mahometans of the mainland of Ceram, which is perhaps due to there having been no indigenes on these small islands when the first settlers arrived. In Ceram, the Alfuros of Papuan race are the predominant type, the Malay physiognomy being seldom well marked; whereas here the reverse is the case, and a slight infusion of Papuan on a mixture of

Malay and Bugis has produced a very good-looking set of people. The lower class of the population consists almost entirely of the indigenes of the adjacent islands. They are a fine race, with strongly-marked Papuan features, frizzly hair, and brown complexions. The Goram language is spoken also at the east end of Ceram, and in the adjacent islands. It has a general resemblance to the languages of Ceram, but possesses a peculiar element which I have not met with in other languages of the Archipelago.

After great delay, considering the importance of every day at this time of year, a miserable boat and five men were found, and with some difficulty I stowed away in it such baggage as it was absolutely necessary for me to take, leaving scarcely sitting or sleeping room. The sailing qualities of the boat were highly vaunted, and I was assured that at this season a small one was much more likely to succeed in making the journey. We first coasted along the island, reaching its eastern extremity the following morning (April 11th), and found a strong W.S.W. wind blowing, which just allowed us to lay across to the Matabello Islands, a distance little short of twenty miles. I did not much like the look of the heavy sky and rather rough sea, and my men were very unwilling to make the attempt; but as we could scarcely hope for a better chance, I insisted upon trying. The pitching and jerking of our little boat

soon reduced me to a state of miserable helplessness, and
I lay down, resigned to whatever might happen. After
three or four hours, I was told we were nearly over; but
when I got up, two hours later, just as the sun was setting,
I found we were still a good distance from the point,
owing to a strong current which had been for some time
against us. Night closed in, and the wind drew more
ahead, so we had to take in sail. Then came a calm, and
we rowed and sailed as occasion offered ; and it was four in
the morning when we reached the village of Kissiwoi, not
having made more than three miles in the last twelve
hours.

MATABELLO ISLANDS.

At daylight I found we were in a beautiful little
harbour, formed by a coral reef about two hundred
yards from shore, and perfectly secure in every wind.
Having eaten nothing since the previous morning, we
cooked our breakfast comfortably on shore, and left
about noon, coasting along the two islands of this group,
which lie in the same line, and are separated by a narrow
channel. Both seem entirely formed of raised coral rock;
but there has been a subsequent subsidence, as shown
by the barrier reef which extends all along them at
varying distances from the shore. This reef is sometimes
only marked by a line of breakers when there is a little

swell on the sea; in other places there is a ridge of dead coral above the water, which is here and there high enough to support a few low bushes. This was the first example I had met with of a true barrier reef due to subsidence, as has been so clearly shown by Mr. Darwin. In a sheltered archipelago they will seldom be distinguishable, from the absence of those huge rolling waves and breakers which in the wide ocean throw up a barrier of broken coral far above the usual high-water mark, while here they rarely rise to the surface.

On reaching the end of the southern island, called Uta, we were kept waiting two days for a wind that would enable us to pass over to the next island, Teor, and I began to despair of ever reaching Ké, and determined on returning. We left with a south wind, which suddenly changed to north-east, and induced me to turn again southward in the hopes that this was the commencement of a few days' favourable weather. We sailed on very well in the direction of Teor for about an hour, after which the wind shifted to W.S.W., and we were driven much out of our course, and at nightfall found ourselves in the open sea, and full ten miles to leeward of our destination. My men were now all very much frightened, for if we went on we might be a week at sea in our little open boat, laden almost to the water's edge; or we might drift on to the coast of New Guinea, in which case we should most likely all be murdered. I could not deny these probabilities, and although I showed them that we could not get back to our starting-point with the wind as it was, they insisted upon returning. We accordingly put about, and found that we could lay no nearer to Uta than to Teor ; however, by great good luck, about ten o'clock we hit upon a little coral island, and lay under its lee till morning, when a favourable change of wind brought us back to Uta, and by evening (April 18th) we reached our first anchorage in Matabello, where I resolved to stay a few days, and then return to Goram. It was with much regret that I gave up my trip to Ké and the intervening islands, which I had looked forward to as likely to make up for my disappointment in Ceram, since my short visit on my voyage to Aru had produced me so many rare and beautiful insects.

The natives of Matabello are almost entirely occupied in making cocoa-nut oil, which they sell to the Bugis and Goram traders, who carry it to Banda and Amboyna. The rugged coral rock seems very favourable to the growth of the cocoa-nut palm, which abounds over the whole island to the very highest points, and produces fruit all the year round. Along with it are great numbers of the areca or betel-nut palm, the nuts of which are sliced, dried, and ground into a paste, which is much used by the betel-chewing Malays and Papuans. All the little children here,

even such as can just run alone, carried between their
lips a mass of the nasty-looking red paste, which is even
more disgusting than to see them at the same age smoking
cigars, which is very common even before they are weaned.
Cocoa-nuts, sweet potatoes, an occasional sago cake, and
the refuse nut after the oil has been extracted by boiling,
form the chief sustenance of these people; and the effect
of this poor and unwholesome diet is seen in the frequency
of eruptions and scurfy skin diseases, and the numerous
sores that disfigure the faces of the children.

The villages are situated on high and rugged coral
peaks, only accessible by steep narrow paths, with ladders
and bridges over yawning chasms. They are filthy with
rotten husks and oil refuse, and the huts are dark, greasy,
and dirty in the extreme. The people are wretched ugly
dirty savages, clothed in unchanged rags, and living in the
most miserable manner, and as every drop of fresh water
has to be brought up from the beach, washing is never
thought of; yet they are actually wealthy, and have the
means of purchasing all the necessaries and luxuries of life.
Fowls are abundant, and eggs were given me whenever I
visited the villages, but these are never eaten, being looked
upon as pets or as merchandise. Almost all of the women
wear massive gold earrings, and in every village there are
dozens of small bronze cannon lying about on the ground,
although they have cost on the average perhaps 10*l.* a-

piece. The chief men of each village came to visit me,
clothed in robes of silk and flowered satin, though their
houses and their daily fare are no better than those of the
other inhabitants. What a contrast between these people
and such savages as the best tribes of hill Dyaks in Borneo,
or the Indians of the Uaupes in South America, living on
the banks of clear streams, clean in their persons and
their houses, with abundance of wholesome food, and
exhibiting its effect in healthy skins and beauty of form
and feature! There is in fact almost as much difference
between the various races of savage as of civilized peoples,
and we may safely affirm that the better specimens of
the former are much superior to the lower examples of
the latter class.

One of the few luxuries of Matabello is the palm wine,
which is the fermented sap from the flower stems of the
cocoa-nut. It is really a very nice drink, more like cyder
than beer, though quite as intoxicating as the latter.
Young cocoa-nuts are also very abundant, so that anywhere
in the island it is only necessary to go a few yards to find
a delicious beverage by climbing up a tree for it. It is
the water of the young fruit that is drunk, before the
pulp has hardened; it is then more abundant, clear, and
refreshing, and the thin coating of gelatinous pulp is
thought a great luxury. The water of full-grown cocoa-
nuts is always thrown away as undrinkable, although it

is delicious in comparison with that of the old dry nuts which alone we obtain in this country. The cocoa-nut pulp I did not like at first; but fruits are so scarce, except at particular seasons, that one soon learns to appreciate anything of a fruity nature.

Many persons in Europe are under the impression that fruits of delicious flavour abound in the tropical forests, and they will no doubt be surprised to learn that the truly wild fruits of this grand and luxuriant archipelago, the vegetation of which will vie with that of any part of the world, are in almost every island inferior in abundance and quality to those of Britain. Wild strawberries and raspberries are found in some places, but they are such poor tasteless things as to be hardly worth eating, and there is nothing to compare with our blackberries and whortleberries. The kanary-nut may be considered equal to a hazel-nut, but I have met with nothing else superior to our crabs, our haws, beech-nuts, wild plums, and acorns; fruits which would be highly esteemed by the natives of these islands, and would form an important part of their sustenance. All the fine tropical fruits are as much cultivated productions as our apples, peaches, and plums, and their wild prototypes, when found, are generally either tasteless or uneatable.

The people of Matabello, like those of most of the Mahometan villages of East Ceram and Goram, amused

me much by their strange ideas concerning the Russian
war. They believe that the Russians were not only most
thoroughly beaten by the Turks, but were absolutely conquered, and all converted to Islamism! And they can
hardly be convinced that such is not the case, and that had
it not been for the assistance of France and England, the
poor Sultan would have fared ill. Another of their
notions is, that the Turks are the largest and strongest
people in the world—in fact a race of giants; that they eat
enormous quantities of meat, and are a most ferocious
and irresistible nation. Whence such strangely incorrect
opinions could have arisen it is difficult to understand,
unless they are derived from Arab priests, or hadjis returned from Mecca, who may have heard of the ancient
prowess of the Turkish armies when they made all Europe
tremble, and suppose that their character and warlike
capacity must be the same at the present time.

GORAM.

A steady south-east wind having set in, we returned to
Manowolko on the 25th of April, and the day after crossed
over to Ondor, the chief village of Goram.

Around this island extends, with few interruptions, an
encircling coral reef about a quarter of a mile from the
shore, visible as a stripe of pale green water, but only at

very lowest ebb-tides showing any rock above the surface. There are several deep entrances through this reef, and inside it there is good anchorage in all weathers. The land rises gradually to a moderate height, and numerous small streams descend on all sides. The mere existence of these streams would prove that the island was not entirely coralline, as in that case all the water would sink through the porous rock as it does at Manowolko and Matabello; but we have more positive proof in the pebbles and stones of their beds, which exhibit a variety of stratified crystalline rocks. About a hundred yards from the beach rises a wall of coral rock, ten or twenty feet high, above which is an undulating surface of rugged coral, which slopes *downward* towards the interior, and then after a slight ascent is bounded by a second wall of coral. Similar walls occur higher up, and coral is found on the highest part of the island.

This peculiar structure teaches us that before the coral was formed land existed in this spot; that this land sunk gradually beneath the waters, but with intervals of rest, during which encircling reefs were formed around it at different elevations; that it then rose to above its present elevation, and is now again sinking. We infer this, because encircling reefs are a proof of subsidence; and if the island were again elevated about a hundred feet, what is now the reef and the shallow sea within it would form a wall of coral rock, and an undulating coralline plain, exactly similar to those that still exist at various altitudes up to the summit of the island. We learn also that these changes have taken place at a comparatively recent epoch, for the surface of the coral has scarcely suffered from the action of the weather, and hundreds of sea-shells, exactly resembling those still found upon the beach, and many of them retaining their gloss and even their colour, are scattered over the surface of the island to near its summit.

Whether the Goram group formed originally part of New Guinea or of Ceram it is scarcely possible to determine, and its productions will throw little light upon the question, if, as I suppose, the islands have been entirely submerged within the epoch of existing species of animals, as in that case it must owe its present fauna and flora to recent immigration from surrounding lands; and with this view its poverty in species very well agrees. It possesses much in common with East Ceram, but at the same time has a good deal of resemblance to the Ké Islands and Banda. The fine pigeon, Carpophaga concinna, inhabits Ké, Banda, Matabello, and Goram, and is replaced by a distinct species, C. neglecta, in Ceram. The insects of these four islands have also a common facies—facts which seem to indicate that some more extensive land has recently disappeared from the area they now occupy,

and has supplied them with a few of its peculiar productions.

The Goram people (among whom 1 stayed a month) are a
race of traders. Every year they visit the Tenimber, Ké,
and Aru Islands, the whole north-west coast of New
Guinea from Oetanata to Salwatty, and the island of
Waigiou and Mysol. They also extend their voyages to
Tidore and Ternate, as well as to Banda and Amboyna.
Their praus are all made by that wonderful race of boat-
builders, the Ké islanders, who annually turn out some
hundreds of boats, large and small, which can hardly be
surpassed for beauty of form and goodness of workmanship.
They trade chiefly in tripang, the medicinal mussoi bark,
wild nutmegs, and tortoise-shell, which they sell to the
Bugis traders at Ceram-laut or Aru, few of them caring to
take their products to any other market. In other respects
they are a lazy race, living very poorly, and much given to
opium smoking. The only native manufactures are sail-
matting, coarse cotton cloth, and pandanus-leaf boxes,
prettily stained and ornamented with shell-work.

In the island of Goram, only eight or ten miles long,
there are about a dozen Rajahs, scarcely better off
than the rest of the inhabitants, and exercising a mere
nominal sway, except when any order is received from
the Dutch Government, when, being backed by a
higher power, they show a little more strict authority

My friend the Rajah of Ammer (commonly called Rajah of Goram) told me that a few years ago, before the Dutch had interfered in the affairs of the island, the trade was not carried on so peaceably as at present, rival praus often fighting when on the way to the same locality, or trafficking in the same village. Now such a thing is never thought of—one of the good effects of the superintendence of a civilized government. Disputes between villages are still, however, sometimes settled by fighting, and I one day saw about fifty men, carrying long guns and heavy cartridge-belts, march through the village. They had come from the other side of the island on some question of trespass or boundary, and were prepared for war if peaceable negotiations should fail.

While at Manowolko I had purchased for 100 florins (9l.) a small prau, which was brought over the next day, as I was informed it was more easy to have the necessary alterations made in Goram, where several Ké workmen were settled.

As soon as we began getting my prau ready I was obliged to give up collecting, as I found that unless I was constantly on the spot myself very little work would be done. As I proposed making some long voyages in this boat, I determined to fit it up conveniently, and was obliged to do all the inside work myself, assisted by my two Amboynese boys. I had plenty of visitors, surprised

to see a white man at work, and much astonished at the
novel arrangements I was making in one of their native
vessels. Luckily I had a few tools of my own, including a
small saw and some chisels, and these were now severely
tried, cutting and fitting heavy iron-wood planks for the
flooring and the posts that support the triangular mast.
Being of the best London make, they stood the work well,
and without them it would have been impossible for me
to have finished my boat with half the neatness, or in
double the time.   I had a Ke workman to put in new ribs,
for which I bought nails of a Bugis trader, at 8*d.* a pound.
My gimlets were, however, too small; and having no augers
we were obliged to bore all the holes with hot irons, a
most tedious and unsatisfactory operation.

Five men had engaged to work at the prau till finished,
and then go with me to Mysol, Waigiou, and Ternate. Their
ideas of work were, however, very different from mine, and
I had immense difficulty with them ; seldom more than
two or three coming together, and a hundred excuses being
given for working only half a day when they did come.
Yet they were constantly begging advances of money,
saying they had nothing to eat. When I gave it them they
were sure to stay away the next day, and when I refused
any further advances some of them declined working any
more.   As the boat approached completion my difficulties
with the men increased.   The uncle of one had commenced

a war, or sort of faction fight, and wanted his assistance;
another's wife was ill, and would not let him come; a
third had fever and ague, and pains in his head and back;
and a fourth had an inexorable creditor who would not let
him go out of his sight. They had all received a month's
wages in advance; and though the amount was not large, it
was necessary to make them pay it back, or I should get
no men at all. I therefore sent the village constable after
two, and kept them in custody a day, when they returned
about three-fourths of what they owed me. The sick man
also paid, and the steersman found a substitute who was
willing to take his debt, and receive only the balance of
his wages.

About this time we had a striking proof of the dangers
of New Guinea trading. Six men arrived at the village
in a small boat almost starved, having escaped out of two
praus, the remainder of whose crews (fourteen in number)
had been murdered by the natives of New Guinea. The
praus had left this village a few months before, and among
the murdered men were the Rajah's son, and the relations
or slaves of many of the inhabitants. The cry of lamen-
tation that arose when the news arrived was most distress-
ing. A score of women, who had lost husbands, brothers,
sons, or more distant relatives, set up at once the most
dismal shrieks and groans and wailings, which continued
at intervals till late at night; and as the chief houses in

the village were crowded together round that which I occupied, our situation was anything but agreeable.

It seems that the village where the attack took place (nearly opposite the small island of Lakahia) is known to be dangerous, and the vessels had only gone there a few days before to buy some tripang. The crew were living on shore, the praus being in a small river close by, and they were attacked and murdered in the day-time while bargaining with the Papuans. The six men who survived were on board the praus, and escaped by at once getting into the small boat and rowing out to sea.

This south-west part of New Guinea, known to the native traders as "Papua Kowiyee" and "Papua Onen," is inhabited by the most treacherous and bloodthirsty tribes. It is in these districts that the commanders and portions of the crews of many of the early discovery ships were murdered, and scarcely a year now passes but some lives are lost. The Goram and Ceram traders are themselves generally inoffensive; they are well acquainted with the character of these natives, and are not likely to provoke an attack by any insults or open attempt at robbery or imposition. They are accustomed to visit the same places every year, and the natives can have no fear of them, as may be alleged in excuse for their attacks on Europeans. In other extensive districts inhabited by the same Papuan races, such as Mysol,

Salwatty, Waigiou, and some parts of the adjacent coast,
the people have taken the first step in civilization, owing
probably to the settlement of traders of mixed breed
among them, and for many years no such attacks have
taken place. On the south-west coast, and in the large
island of Jobie, however, the natives are in a very barbarous condition, and take every opportunity of robbery
and murder,—a habit which is confirmed by the impunity
they experience, owing to the vast extent of wild mountain
and forest country forbidding all pursuit or attempt at
punishment. In the very same village, four years before,
more than fifty Goram men were murdered; and as
these savages obtain an immense booty in the praus
and all their appurtenances, it is to be feared that such
attacks will continue to be made at intervals as long as
traders visit the same spots and attempt no retaliation.
Punishment could only be inflicted on these people by
very arbitrary measures, such as by obtaining possession
of some of the chiefs by stratagem, and rendering them
responsible for the capture of the murderers at the peril of
their own heads. But anything of this kind would be
quite contrary to the system adopted by the Dutch
Government in its dealings with natives.

GORAM TO WAHAI IN CERAM.

When my boat was at length launched and loaded, I got
my men together, and actually set sail the next day (May
27th), much to the astonishment of the Goram people, to
whom such punctuality was a novelty.   I had a crew of
three men and a boy, besides my two Amboyna lads;
which was sufficient for sailing, though rather too few if
obliged to row much.   The next day was very wet,
with squalls, calms, and contrary winds, and with some
difficulty we reached Kilwaru, the metropolis of the Bugis
traders in the far East.   As I wanted to make some
purchases, I stayed here two days, and sent two of my
boxes of specimens by a Macassar prau to be forwarded to
Ternate, thus relieving myself of a considerable incum-
brance.   I bought knives, basins, and handkerchiefs for
barter, which with the choppers, cloth, and beads I had
brought with me, made a pretty good assortment.   I also
bought two tower muskets to satisfy my crew, who insisted
on the necessity of being armed against attacks of pirates ;
and with spices and a few articles of food for the voyage
nearly my last doit was expended.

The little island of Kilwaru is a mere sandbank, just
large enough to contain a small village, and situated
between the islands of Ceram-laut, and Kissa—straits about

a third of a mile wide separating it from each of them.
It is surrounded by coral reefs, and offers good anchorage
in both monsoons. Though not more than fifty yards
across, and not elevated more than three or four feet above
the highest tides, it has wells of excellent drinking water—
a singular phenomenon, which would seem to imply deep-
seated subterranean channels connecting it with other
islands. These advantages, with its situation in the centre
of the Papuan trading district, lead to its being so much
frequented by the Bugis traders. Here the Goram men
bring the produce of their little voyages, which they exchange for cloth, sago cakes, and opium ; and the inhabitants of all the surrounding islands visit it with the
same object. It is the rendezvous of the praus trading to
various parts of New Guinea, which here assort and dry
their cargoes, and refit for the voyage home. Tripang and
mussoi bark are the most bulky articles of produce
brought here, with wild nutmegs, tortoise-shell, pearls, and
birds of Paradise. in smaller quantities. The villagers of
the mainland of Ceram bring their sago, which is thus
distributed to the islands farther east, while rice from
Bali and Macassar can also be purchased at a moderate
price. The Goram men come here for their supplies of
opium, both for their own consumption and for barter in
Mysol and Waigiou, where they have introduced it, and
where the chiefs and wealthy men are passionately fond of

it. Schooners from Bali come to buy Papuan slaves, while the sea-wandering Bugis arrive from distant Singapore in their lumbering praus, bringing thence the produce of the Chinamen's workshops and Kling's bazaar, as well as of the looms of Lancashire and Massachusetts.

One of the Bugis traders who had arrived a few days before from Mysol, brought me news of my assistant Charles Allen, with whom he was well acquainted, and who, he assured me, was making large collections of birds and insects, although he had not obtained any birds of Paradise; Silinta, where he was staying, not being a good place for them. This was on the whole satisfactory, and I was anxious to reach him as soon as possible.

Leaving Kilwaru early in the morning of June 1st, with a strong east wind we doubled the point of Ceram about noon, the heavy sea causing my prau to roll about a good deal, to the damage of our crockery. As bad weather seemed coming on, we got inside the reefs and anchored opposite the village of Warus-warus to wait for a change. The night was very squally, and though in a good harbour we rolled and jerked uneasily; but in the morning I had greater cause for uneasiness in the discovery that our entire Goram crew had decamped, taking with them all they possessed and a little more, and leaving us without any small boat in which to land. I immediately told my Amboyna men to load and fire the muskets as a signal of

distress, which was soon answered by the village chief
sending off a boat, which took me on shore. I requested
that messengers should be immediately sent to the neighbouring villages in quest of the fugitives, which was
promptly done. My prau was brought into a small creek,
where it could securely rest in the mud at low water, and
part of a house was given me in which I could stay for
a while. I now found my progress again suddenly checked,
just when I thought I had overcome my chief difficulties.
As I had treated my men with the greatest kindness, and
had given them almost everything they had asked for, I
can impute their running away only to their being totally
unaccustomed to the restraint of a European master, and
to some undefined dread of my ultimate intentions regard-
ing them. The oldest man was an opium smoker, and a
reputed thief, but I had been obliged to take him at the
last moment as a substitute for another. I feel sure it was
he who induced the others to run away, and as they knew
the country well, and had several hours' start of us, there
was little chance of catching them.

We were here in the great sago district of East Ceram,
which supplies most of the surrounding islands with their
daily bread, and during our week's delay I had an opportunity of seeing the whole process of making it, and
obtaining some interesting statistics. The sago tree is a
palm, thicker and larger than the cocoa-nut tree, although

rarely so tall, and having immense pinnate spiny leaves,
which completely cover the trunk till it is many years old.
It has a creeping root-stem like the Nipa palm, and when
about ten or fifteen years of age sends up an immense
terminal spike of flowers, after which the tree dies. It
grows in swamps, or in swampy hollows on the rocky
slopes of hills, where it seems to thrive equally well as
when exposed to the influx of salt or brackish water.
The midribs of the immense leaves form one of the most
useful articles in these lands, supplying the place of
bamboo, to which for many purposes they are superior.
They are twelve or fifteen feet long, and, when very fine,
as thick in the lower part as a man's leg. They are very
light, consisting entirely of a firm pith covered with a hard
thin rind or bark. Entire houses are built of these; they
form admirable roofing-poles for thatch; split and well-
supported, they do for flooring; and when chosen of equal
size, and pegged together side by side to fill up the panels
of framed wooden houses, they have a very neat appearance, and make better walls and partitions than boards, as
they do not shrink, require no paint or varnish, and are
not a quarter the expense. When carefully split and
shaved smooth they are formed into light boards with pegs
of the bark itself, and are the foundation of the leaf-
covered boxes of Goram. All the insect-boxes I used in
the Moluccas were thus made at Amboyna, and when

covered with stout paper inside and out, are strong, light, and secure the insect-pins remarkably well. The leaflets of the sago folded and tied side by side on the smaller midribs form the "atap" or thatch in universal use, while the product of the trunk is the staple food of some hundred thousands of men.

When sago is to be made, a full-grown tree is selected just before it is going to flower. It is cut down close to the ground, the leaves and leaf-stalks cleared away, and a broad strip of the bark taken off the upper side of the trunk. This exposes the pithy matter, which is of a rusty colour near the bottom of the tree, but higher up pure white, about as hard as a dry apple, but with woody fibres running through it about a quarter of an inch apart. This pith is cut or broken down into a coarse powder by means of a tool constructed for the purpose—a club of hard and heavy wood, having a piece of sharp quartz rock firmly imbedded into its blunt end, and projecting about half an

T.B.

SAGO CLUB.

inch. By successive blows of this, narrow strips of the pith are cut away, and fall down into the cylinder formed by the bark. Proceeding steadily on, the whole trunk is cleared out, leaving a skin not more than half an inch in thickness. This material is carried away (in baskets made of the sheathing bases of the leaves) to the nearest water, where a washing-machine is put up, which is composed

SAGO WASHING.

almost entirely of the sago tree itself. The large sheathing bases of the leaves form the troughs, and the fibrous covering from the leaf-stalks of the young cocoa-nut the strainer. Water is poured on the mass of pith, which is kneaded and pressed against the strainer till the starch is all dissolved and has passed through, when the fibrous refuse is thrown away, and a fresh basketful put in its place. The water

charged with sago starch passes on to a trough, with a
depression in the centre, where the sediment is deposited,
the surplus water trickling off by a shallow outlet. When
the trough is nearly full, the mass of starch, which has a
slight reddish tinge, is made into cylinders of about thirty
pounds' weight, and neatly covered with sago leaves, and
in this state is sold as raw sago.

Boiled with water this forms a thick glutinous mass,
with a rather astringent taste, and is eaten with salt,
limes, and chilies. Sago-bread is made in large quantities, by baking it into cakes in a small clay oven
containing six or eight slits side by side, each about
three-quarters of an inch wide, and six or eight inches
square. The raw sago is broken up, dried in the sun,
powdered, and finely sifted. The oven is heated over a
clear fire of embers, and is lightly filled with the sago-
powder. The openings are then covered with a flat piece

SAGO OVEN.

of sago bark, and in about
five minutes the cakes are
turned out sufficiently baked.
The hot cakes are very nice
with butter, and when made
with the addition of a little
sugar and grated cocoa-nut
are quite a delicacy. They are soft, and something like
corn-flour cakes, but have a slight characteristic flavour

which is lost in the refined sago we use in this country. When not wanted for immediate use, they are dried for several days in the sun, and tied up in bundles of twenty. They will then keep for years; they are very hard, and very rough and dry, but the people are used to them from infancy, and little children may be seen gnawing at them as contentedly as ours with their bread-and-butter. If dipped in water and then toasted, they become almost as good as when fresh baked; and thus treated they were my daily substitute for bread with my coffee. Soaked and boiled they make a very good pudding or vegetable, and served well to economize our rice, which is sometimes difficult to get so far east.

It is truly an extraordinary sight to witness a whole tree-trunk, perhaps twenty feet long and four or five in circumference, converted into food with so little labour and preparation. A good-sized tree will produce thirty tomans or bundles of thirty pounds each, and each toman will make sixty cakes of three to the pound. Two of these cakes are as much as a man can eat at one meal, and five are considered a full day's allowance; so that, reckoning a tree to produce 1,800 cakes, weighing 600 pounds, it will supply a man with food for a whole year. The labour to produce this is very moderate. Two men will finish a tree in five days, and two women will bake the whole into cakes in five days more; but the raw sago will keep very

well, and can be baked as wanted, so that we may estimate
that in ten days a man may produce food for the whole
year. This is on the supposition that he possesses sago
trees of his own, for they are now all private property. If
he does not, he has to pay about seven and sixpence for
one ; and as labour here is five pence a day, the total cost
of a year's food for one man is about twelve shillings.
The effect of this cheapness of food is decidedly prejudicial,
for the inhabitants of the sago countries are never so well
off as those where rice is cultivated. Many of the people
here have neither vegetables nor fruit, but live almost
entirely on sago and a little fish. Having few occupations
at home, they wander about on petty trading or fishing
expeditions to the neighbouring islands ; and as far as the
comforts of life are concerned, are much inferior to the
wild hill-Dyaks of Borneo, or to many of the more barbarous tribes of the Archipelago.

The country round Warus-warus is low and swampy,
and owing to the absence of cultivation there were scarcely
any paths leading into the forest. I was therefore unable
to collect much during my enforced stay, and found no
rare birds or insects to improve my opinion of Ceram as a
collecting ground. Finding it quite impossible to get men
here to accompany me on the whole voyage, I was obliged
to be content with a crew to take me as far as Wahai, on
the middle of the north coast of Ceram, and the chief

Dutch station in the island. The journey took us five
days, owing to calms and light winds, and no incident of
any interest occurred on it, nor did I obtain at our
stopping places a single addition to my collections worth
naming. At Wahai, which I reached on the 15th of June,
I was hospitably received by the Commandant and my old
friend Herr Rosenberg, who was now on an official visit
here. He lent me some money to pay my men, and I was
lucky enough to obtain three others willing to make the
voyage with me to Ternate, and one more who was to
return from Mysol. One of my Amboyna lads, however,
left me, so that I was still rather short of hands.

I found here a letter from Charles Allen, who was at
Silinta in Mysol, anxiously expecting me, as he was out of
rice and other necessaries, and was short of insect-pins. He
was also ill, and if I did not soon come would return to
Wahai.

As my voyage from this place to Waigiou was among
islands inhabited by the Papuan race, and was an event-
ful and disastrous one, I will narrate its chief inci-
dents in a separate chapter in that division of my work
devoted to the Papuan Islands. I now have to pass over
a year spent in Waigiou and Timor, in order to describe
my visit to the island of Bouru, which concluded my
explorations of the Moluccas.

CHAPTER XXVI.

BOURU.

(MAY AND JUNE 1861. *Map*, p. 74.)

I HAD long wished to visit the large island of Bouru, which lies due west of Ceram, and of which scarcely anything appeared to be known to naturalists, except that it contained a babirusa very like that of Celebes. I therefore made arrangements for staying there two months after leaving Timor Delli in 1861. This I could conveniently do by means of the Dutch mail-steamers, which make a monthly round of the Moluccas.

We arrived at the harbour of Cajeli on the 4th of May; a gun was fired, the Commandant of the fort came along-side in a native boat to receive the post-packet, and took me and my baggage on shore, the steamer going off again without coming to an anchor. We went to the house of the Opzeiner, or overseer, a native of Amboyna—Bouru being too poor a place to deserve even an Assistant Resident; yet the appearance of the village was very far superior to that of Delli, which possesses " His Excellency the Governor," and the little fort, in perfect order, surrounded by neat grass-plots and straight walks, although manned by only a dozen Javanese soldiers with an Adjutant for commander, was a very Sebastopol in comparison with the miserable mud enclosure at Delli, with its numerous staff of Lieutenants, Captain, and Major. Yet this, as well as most of the forts in the Moluccas, was originally built by the Portuguese themselves. Oh! Lusitania, how art thou fallen!

While the Opzeiner was reading his letters, I took a walk round the village with a guide in search of a house. The whole place was dreadfully damp and muddy, being built in a swamp with not a spot of ground raised a foot above it, and surrounded by swamps on every side. The houses were mostly well built, of wooden framework filled in with gaba-gaba (leaf-stems of the sago-palm), but as they had no whitewash, and the floors were of bare black earth like the roads, and generally on the same level, they were extremely damp and gloomy. At length I found one with the floor raised about a foot, and succeeded in making a bargain with the owner to turn out immediately, so that by night I had installed myself comfortably. The chairs and tables were left for me; and as the whole of the remaining furniture in the house consisted of a little crockery and a few clothes-boxes, it was not much trouble for the owners to move into the house of some relatives,

and thus obtain a few silver rupees very easily. Every
foot of ground between the houses throughout the village
is crammed with fruit trees, so that the sun and air have
no chance of penetrating. This must be very cool and
pleasant in the dry season, but makes it damp and un-
healthy at other times of the year. Unfortunately I had
come two months too soon, for the rains were not yet over,
and mud and water were the prominent features of the
country.

About a mile behind and to the east of the village the
hills commence, but they are very barren, being covered
with scanty coarse grass and scattered trees of the
Melaleuca cajuputi, from the leaves of which the celebrated cajeput oil is made. Such districts are absolutely
destitute of interest for the zoologist. A few miles further
on rose higher mountains, apparently well covered with
forest, but they were entirely uninhabited and trackless,
and practically inaccessible to a traveller with limited
time and means. It became evident, therefore, that I
must leave Cajeli for some better collecting ground, and
finding a man who was going a few miles eastward to a
village on the coast where he said there were hills and
forest, I sent my boy Ali with him to explore and report
on the capabilities of the district. At the same time I
arranged to go myself on a little excursion up a river
which flows into the bay about five miles north of the

town, to a village of the Alfuros, or indigenes, where I
thought I might perhaps find a good collecting ground.

The Rajah of Cajeli, a good-tempered old man, offered to
accompany me, as the village was under his government;
and we started one morning early, in a long narrow boat
with eight rowers. In about two hours we entered the
river, and commenced our inland journey against a very
powerful current. The stream was about a hundred yards
wide, and was generally bordered with high grass, and
occasionally bushes and palm-trees. The country round
was flat and more or less swampy, with scattered trees and
shrubs. At every bend we crossed the river to avoid the
strength of the current, and arrived at our landing-
place about four o'clock, in a torrent of rain. Here we
waited for an hour, crouching under a leaky mat till
the Alfuros arrived who had been sent for from the
village to carry my baggage, when we set off along a
path of whose extreme muddiness I had been warned
before starting.

I turned up my trousers as high as possible, grasped a
stout stick to prevent awkward falls, and then boldly
plunged into the first mud-hole, which was immediately
succeeded by another and another. The mud or mud and
water was knee-deep, with little intervals of firmer ground
between, making progression exceedingly difficult. The
path was bordered with high rigid grass, growing in dense

clumps separated by water, so that nothing was to be gained by leaving the beaten track, and we were obliged to go floundering on, never knowing where our feet would rest, as the mud was now a few inches, now two feet, deep, and the bottom very uneven, so that the foot slid down to the lowest part, and made it difficult to keep one's balance. One step would be upon a concealed stick or log, almost dislocating the ankle, while the next would plunge into soft mud above the knee. It rained all the way, and the long grass, six feet high, met over the path ; so that we could not see a step of the way ahead, and received a double drenching. Before we got to the village it was dark, and we had to cross over a small but deep and swollen stream by a narrow log of wood, which was more than a foot under water. There was a slender shaking stick for a handrail, and it was nervous work feeling in the dark in the rushing water for a safe place on which to place the advanced foot. After an hour of this most disagreeable and fatiguing walk we reached the village, followed by the men with our guns, ammunition, boxes, and bedding, all more or less soaked. We consoled ourselves with some hot tea and cold fowl, and went early to bed.

The next morning was clear and fine, and I set out soon after sunrise to explore the neighbourhood. The village had evidently been newly formed, and consisted of a single straight street of very miserable huts totally deficient in every comfort, and as bare and cheerless inside as out. It was situated on a little elevated patch of coarse gravelly soil, covered with the usual high rigid grass, which came up close to the backs of the houses. At a short distance in several directions were patches of forest, but all on low and swampy ground. I made one attempt along the only path I could find, but soon came upon a deep mud-hole, and found that I must walk barefoot if at all ; so I returned and deferred further exploration till after breakfast. I then went on into the jungle and found patches of sago-palms and a low forest vegetation, but the paths were every-where full of mud-holes, and intersected by muddy streams and tracts of swamp, so that walking was not pleasurable, and too much attention to one's steps was not favourable to insect catching, which requires above everything freedom of motion. I shot a few birds, and caught a few butterflies, but all were the same as I had already obtained about Cajeli.

On my return to the village I was told that the same kind of ground extended for many miles in every direction, and I at once decided that Wayapo was not a suitable place to stay at. The next morning early we waded back again through the mud and long wet grass to our boat, and by mid-day reached Cajeli, where I waited Ali's return to decide on my future movements. He came the following day, and gave a very bad account of Pelah, where he had been. There was

a little brush and trees along the beach, and hills inland
covered with high grass and cajuputi trees—my dread and
abhorrence. On inquiring who could give me trustworthy
information, I was referred to the Lieutenant of the
Burghers, who had travelled all round the island, and was a
very intelligent fellow. I asked him to tell me if he knew
of any part of Bouru where there was no " kusu-kusu," as
the coarse grass of the country is called. He assured me
that a good deal of the south coast was forest land, while
along the north was almost entirely swamp and grassy hills.
After minute inquiries, I found that the forest country commenced at a place called Waypoti, only a few miles beyond
Pelah, but that, as the coast beyond that place was exposed
to the east monsoon and dangerous for praus, it was necessary to walk. I immediately went to the Opzeiner, and
he called the Rajah. We had a consultation, and arranged
for a boat to take me the next evening but one, to
Pelah, whence I was to proceed on foot, the Orang-kaya
going the day before to call the Alfuros to carry my
baggage.

The journey was made as arranged, and on May 19th
we arrived at Waypoti, having walked about ten miles
along the beach, and through stony forest bordering the
sea, with occasional plunges of a mile or two into the
interior. We found no village, but scattered houses and
plantations, with hilly country pretty well covered with

forest, and looking rather promising. A low hut with a very rotten roof, showing the sky through in several places, was the only one I could obtain. Luckily it did not rain that night, and the next day we pulled down some of the walls to repair the roof, which was of immediate importance, especially over our beds and table.

About half a mile from the house was a fine mountain stream, running swiftly over a bed of rocks and pebbles, and beyond this was a hill covered with fine forest. By carefully picking my way I could wade across this river without getting much above my knees, although I would sometimes slip off a rock and go into a hole up to my waist, and about twice a week I went across it in order to explore the forest. Unfortunately there were no paths here of any extent, and it did not prove very productive either in insects or birds. To add to my difficulties I had stupidly left my only pair of strong boots on board the steamer, and my others were by this time all dropping to pieces, so that I was obliged to walk about barefooted, and in constant fear of hurting my feet, and causing a wound which might lay me up for weeks, as had happened in Borneo, Aru, and Dorey. Although there were numerous plantations of maize and plantains, there were no new clearings; and as without these it is almost impossible to find many of the best kinds of insects, I determined to make one myself, and with much difficulty engaged two

K 2

men to clear a patch of forest, from which I hoped to obtain many fine beetles before I left.

During the whole of my stay, however, insects never became plentiful. My clearing produced me a few fine longicorns and Buprestidæ, different from any I had before seen, together with several of the Amboyna species, but by no means so numerous or so beautiful as I had found in that small island. For example, I collected only 210 different kinds of beetles during my two months' stay at Bouru, while in three weeks at Amboyna, in 1857, I found more than 300 species. One of the finest insects found at Bouru was a large Cerambyx, of a deep shining chestnut colour, and with very long antennæ. It varied greatly in size, the largest specimens being three inches long, while the smallest were only an inch, the antennæ varying from one and a half to five inches.

One day my boy Ali came home with a story of a big snake. He was walking through some high grass, and stepped on something which he took for a small fallen tree, but it felt cold and yielding to his feet, and far to the right and left there was a waving and rustling of the herbage. He jumped back in affright and prepared to shoot, but could not get a good view of the creature, and it passed away, he said, like a tree being dragged along through the grass. As he had several times already shot large snakes, which he declared were all as nothing compared with this, I am inclined to believe it must really have been a monster. Such creatures are rather plentiful here, for a man living close by showed me on his thigh the marks where he had been seized by one close to his house. It was big enough to take the man's thigh in its mouth, and he would pro- bably have been killed and devoured by it had not his cries brought out his neighbours, who destroyed it with their choppers. As far as I could make out it was about twenty feet long, but Ali's was probably much larger.

It sometimes amuses me to observe how, a few days after I have taken possession of it, a native hut seems quite a comfortable home. My house at Waypoti was a bare shed, with a large bamboo platform at one side. At one end of this platform, which was elevated about three feet, I fixed up my mosquito curtain, and partly enclosed it with a large Scotch plaid, making a comfortable little sleeping apartment. I put up a rude table on legs buried in the earthen floor, and had my comfortable rattan-chair for a seat. A line across one corner carried my daily- washed cotton clothing, and on a bamboo shelf was arranged my small stock of crockery and hardware. Boxes were ranged against the thatch walls, and hanging shelves, to preserve my collections from ants while drying, were suspended both without and within the house. On my table lay books, penknives, scissors, pliers, and pins, with

insect and bird labels, all of which were unsolved mysteries
to the native mind.

Most of the people here had never seen a pin, and
the better informed took a pride in teaching their more
ignorant companions the peculiarities and uses of that
strange European production—a needle with a head, but
no eye ! Even paper, which we throw away hourly as
rubbish, was to them a curiosity ; and I often saw them
picking up little scraps which had been swept out of
the house, and carefully putting them away in their betel-
pouch. Then when I took my morning coffee and evening
tea, how many were the strange things displayed to them!
Teapot, teacups, teaspoons, were all more or less curious in
their eyes ; tea, sugar, biscuit, and butter, were articles of
human consumption seen by many of them for the first
time. One asks if that whitish powder is " gula passir "
(sand-sugar), so called to distinguish it from the coarse
lump palm-sugar or molasses of native manufacture ; and
the biscuit is considered a sort of European sago-cake,
which the inhabitants of those remote regions are obliged
to use in the absence of the genuine article. My pursuits
were of course utterly beyond their comprehension. They
continually asked me what white people did with the birds
and insects I took so much care to preserve. If I only
kept what was beautiful, they might perhaps comprehend
it ; but to see ants and flies and small ugly insects put

away so carefully was a great puzzle to them, and they were convinced that there must be some medical or magical use for them which I kept a profound secret. These people were in fact as completely unacquainted with civilized life as the Indians of the Rocky Mountains, or the savages of Central Africa—yet a steamship, that highest triumph of human ingenuity, with its little floating epitome of European civilization, touches monthly at Cajeli, twenty miles off; while at Amboyna, only sixty miles distant, a European population and government have been established for more than three hundred years.

Having seen a good many of the natives of Bouru from different villages, and from distant parts of the island, I feel convinced that they consist of two distinct races now partially amalgamated. The larger portion are Malays of the Celebes type, often exactly similar to the Tomóre people of East Celebes, whom I found settled in Batchian ; while others altogether resemble the Alfuros of Ceram. The influx of two races can easily be accounted for. The Sula Islands, which are closely connected with East Celebes, approach to within forty miles of the north coast of Bouru, while the island of Manipa offers an easy point of departure for the people of Ceram. I was confirmed in this view by finding that the languages of Bouru possessed distinct resemblances to that of Sula, as well as to those of Ceram.

Soon after we had arrived at Waypoti, Ali had seen a beautiful little bird of the genus Pitta, which I was very anxious to obtain, as in almost every island the species are different, and none were yet known from Bouru. He and my other hunter continued to see it two or three times a week, and to hear its peculiar note much oftener, but could never get a specimen, owing to its always frequenting the most dense thorny thickets, where only hasty glimpses of it could be obtained, and at so short a distance that it would be difficult to avoid blowing the bird to pieces. Ali was very much annoyed that he could not get a specimen of this bird, in going after which he had already severely wounded his feet with thorns; and when we had only two days more to stay, he went of his own accord one evening to sleep at a little hut in the forest some miles off, in order to have a last try for it at daybreak, when many birds come out to feed, and are very intent on their morning meal. The next evening he brought me home two specimens, one with the head blown completely off, and otherwise too much injured to preserve, the other in very good order, and which I at once saw to be a new species, very like the Pitta celebensis, but ornamented with a square patch of bright red on the nape of the neck.

The next day after securing this prize we returned to Cajeli, and packing up my collections left Bouru by the steamer. During our two days' stay at Ternate, I took on board what baggage I had left there, and bade adieu to all my friends. We then crossed over to Menado, on our way to Macassar and Java, and I finally quitted the Moluccas, among whose luxuriant and beautiful islands I had wandered for more than three years.

My collections in Bouru, though not extensive, were of considerable interest; for out of sixty-six species of birds which I collected there, no less than seventeen were new, or had not been previously found in any island of the Moluccas. Among these were two kingfishers, Tanysiptera acis and Ceyx Cajeli; a beautiful sunbird, Nectarinea proserpina; a handsome little black and white flycatcher, Monarcha loricata, whose swelling throat was beautifully scaled with metallic blue; and several of less interest. I also obtained a skull of the babirusa, one specimen of which was killed by native hunters during my residence at Cajeli.

CHAPTER XXVII.

THE Moluccas consist of three large islands, Gilolo, Ceram, and Bouru, the two former being each about two hundred miles long ; and a great number of smaller isles and islets, the most important of which are Batchian, Morty, Obi, Ké, Timor-laut, and Amboyna; and among the smaller ones, Ternate, Tidore, Kaióa, and Banda. These occupy a space of ten degrees of latitude by eight of longitude, and they are connected by groups of small islets to New Guinea on the east, the Philippines on the north, Celebes on the west, and Timor on the south. It will be as well to bear in mind these main features of extent and geographical position, while we survey their animal productions and discuss their relations to the countries which surround them on every side in almost equal proximity.

We will first consider the Mammalia, or warm-blooded quadrupeds, which present us with some singular anomalies. The land mammals are exceedingly few in number, only

ten being yet known from the entire group. The bats or
aërial mammals, on the other hand, are numerous—not less
than twenty-five species being already known. But even
this exceeding poverty of terrestrial mammals does not at
all represent the real poverty of the Moluccas in this
class of animals; for, as we shall soon see, there is good
reason to believe that several of the species have been
introduced by man, either purposely or by accident.

The only quadrumanous animal in the group is the
curious baboon-monkey, Cynopithecus nigrescens, already
described as being one of the characteristic animals of
Celebes. This is found only in the island of Batchian;
and it seems so much out of place there—as it is difficult
to imagine how it could have reached the island by any
natural means of dispersal, and yet not have passed by
the same means over the narrow strait to Gilolo—that
it seems more likely to have originated from some individuals which had escaped from confinement, these and
similar animals being often kept as pets by the Malays,
and carried about in their praus.

Of all the carnivorous animals of the Archipelago the
only one found in the Moluccas is the Viverra tangalunga,
which inhabits both Batchian and Bouru, and probably
some of the other islands. I am inclined to think that
this also may have been introduced accidentally, for it is
often made captive by the Malays, who procure civet

from it, and it is an animal very restless and untameable, and therefore likely to escape. This view is rendered still more probable by what Antonio de Morga tells us was the custom in the Philippines in 1602. He says that "the natives of Mindanao carry about civet-cats in cages, and sell them in the islands; and they take the civet from them, and let them go again." The same species is common in the Philippines and in all the large islands of the Indo-Malay region.

The only Moluccan ruminant is a deer, which was once supposed to be a distinct species, but is now generally considered to be a slight variety of the Rusa hippelaphus of Java. Deer are often tamed and petted, and their flesh is so much esteemed by all Malays, that it is very natural they should endeavour to introduce them into the remote islands in which they settled, and whose luxuriant forests seem so well adapted for their subsistence.

The strange babirusa of Celebes is also found in Bouru, but in no other Moluccan island, and it is somewhat difficult to imagine how it got there. It is true that there is some approximation between the birds of the Sula Islands (where the babirusa is also found) and those of Bouru, which seems to indicate that these islands have recently been closer together, or that some intervening land has disappeared. At this time the babirusa may have entered Bouru, since it probably swims as well as its allies the pigs. These are spread all over the Archipelago, even to several of the smaller islands, and in many cases the species are peculiar. It is evident, therefore, that they have some natural means of dispersal. There is a popular idea that pigs cannot swim, but Sir Charles Lyell has shown that this is a mistake. In his "Principles of Geology" (10th Edit. vol. ii. p. 355) he adduces evidence to show that pigs have swum many miles at sea, and are able to swim with great ease and swiftness. I have myself seen a wild pig swimming across the arm of the sea that separates Singapore from the Peninsula of Malacca, and we thus have explained the curious fact, that of all the large mammals of the Indian region, pigs alone extend beyond the Moluccas and as far as New Guinea, although it is somewhat curious that they have not found their way to Australia.

The little shrew, Sorex myosurus, which is common in Sumatra, Borneo, and Java, is also found in the larger islands of the Moluccas, to which it may have been accidentally conveyed in native praus.

This completes the list of the placental mammals which are so characteristic of the Indian region; and we see that, with the single exception of the pig, all may very probably have been introduced by man, since all except the pig are of species identical with those now abounding in the great Malay islands, or in Celebes.

The four remaining mammals are Marsupials, an order of the class Mammalia, which is very characteristic of the Australian fauna; and these are probably true natives of the Moluccas, since they are either of peculiar species, or if found elsewhere are natives only of New Guinea or North Australia. The first is the small flying opossum, Belideus ariel, a beautiful little animal, exactly like a small flying squirrel in appearance, but belonging to the marsupial order. The other three are species of the curious genus Cuscus, which is peculiar to the Austro-Malayan region. These are opossum-like animals, with a long prehensile tail, of which the terminal half is generally bare. They have small heads, large eyes, and a dense covering of woolly fur, which is often pure white with irregular black spots or blotches, or sometimes ashy brown with or without white spots. They live in trees, feeding upon the leaves, of which they devour large quantities. They move about slowly, and are difficult to kill, owing to the thickness of their fur, and their tenacity of life. A heavy charge of shot will often lodge in the skin and do them no harm, and even breaking the spine or piercing the brain will not kill them for some hours. The natives everywhere eat their flesh, and as their motions are so slow, easily catch them by climbing; so that it is wonderful they have not been exterminated. It may be, however, that their dense woolly fur protects them from birds of prey, and the islands they live in are too thinly inhabited for man to be able to exterminate them. The figure represents Cuscus ornatus, a new species discovered by me in Batchian, and

CUSCUS ORNATUS.

which also inhabits Ternate. It is peculiar to the Moluccas, while the two other species which inhabit Ceram are found also in New Guinea and Waigiou.

In place of the excessive poverty of mammals which characterises the Moluccas, we have a very rich display of the feathered tribes. The number of species of birds at present known from the various islands of the Moluccan group is 265, but of these only 70 belong to the usually abundant tribes of the waders and swimmers, indicating that these are very imperfectly known. As they are also pre-eminently wanderers, and are thus little fitted for illustrating the geographical distribution of life in a limited area, we will here leave them out of consideration and confine our attention only to the 195 land birds.

When we consider that all Europe, with its varied climate and vegetation, with every mile of its surface explored, and with the immense extent of temperate Asia and Africa, which serve as storehouses, from which it is continually recruited, only supports 257 species of land birds as residents or regular immigrants, we must look upon the numbers already procured in the small and comparatively unknown islands of the Moluccas as indicating a fauna of fully average richness in this department. But when we come to examine the family groups which go to make up this number, we find the most curious deficiencies in some, balanced by equally striking redundancy in others. Thus if we compare the birds of the Moluccas with those of India, as given in Mr. Jerdon's work, we find that the three groups of the parrots, kingfishers, and pigeons, form nearly *one-third* of the whole land-birds in the former, while they amount to only *one-twentieth* in the latter country. On the other hand, such wide-spread groups as the thrushes, warblers, and finches, which in India form nearly *one-third* of all the land-birds, dwindle down in the Moluccas to *one-fourteenth.*

The reason of these peculiarities appears to be, that the Moluccan fauna has been almost entirely derived from that of New Guinea, in which country the same deficiency and the same luxuriance is to be observed. Out of the seventy-eight genera in which the Moluccan land-birds may be classed, no less than seventy are characteristic of New Guinea, while only six belong specially to the Indo-Malay islands. But this close resemblance to New Guinea genera does not extend to the species, for no less than 140 out of the 195 land-birds are peculiar to the Moluccan islands, while 32 are found also in New Guinea, and 15 in the Indo-Malay islands. These facts teach us, that though the birds of this group have evidently been derived mainly from New Guinea, yet the immigration has not been a recent one, since there has been time for the greater portion of the species to have become changed. We find, also, that many very characteristic New Guinea forms have not entered the Moluccas at all, while others found in Ceram and Gilolo do not extend so far west as Bouru. Considering, further, the absence of most of the New Guinea mammals from the Moluccas, we are led to the conclusion that these islands are not fragments which have been separated from New Guinea, but form a distinct insular region, which has been upheaved independently at a rather remote epoch, and during all the mutations it has undergone has been constantly receiving immigrants from that great and productive island. The considerable length of time the Moluccas have remained isolated is further indicated by the occurrence of two peculiar genera of birds, Semioptera and Lycocorax, which are found nowhere else.

We are able to divide this small archipelago into two well-marked groups—that of Ceram, including also Bouru, Amboyna, Banda, and Ké; and that of Gilolo, including Morty, Batchian, Obi, Ternate, and other small islands. These divisions have each a considerable number of peculiar species, no less than fifty-five being found in the Ceram group only; and besides this, most of the separate islands have some species peculiar to themselves. Thus Morty island has a peculiar kingfisher, honeysucker, and starling; Ternate has a ground-thrush (Pitta) and a fly-catcher; Banda has a pigeon, a shrike, and a Pitta; Ké has two flycatchers, a Zosterops, a shrike, a king-crow, and a cuckoo; and the remote Timor-laut, which should probably come into the Moluccan group, has a cockatoo and lory as its only known birds, and both are of peculiar species.

The Moluccas are especially rich in the parrot tribe, no

less than twenty-two species, belonging to ten genera, inhabiting them. Among these is the large red-crested cockatoo, so commonly seen alive in Europe, two handsome red parrots of the genus Eclectus, and five of the beautiful crimson lories, which are almost exclusively confined to these islands and the New Guinea group. The pigeons are hardly less abundant or beautiful, twenty-one species being known, including twelve of the beautiful green fruit pigeons, the smaller kinds of which are ornamented with the most brilliant patches of colour on the head and the under-surface. Next to these come the kingfishers, including sixteen species, almost all of which are beautiful, and many are among the most brilliantly-coloured birds that exist.

One of the most curious groups of birds, the Megapodii, or mound-makers, is very abundant in the Moluccas. They are gallinaceous birds, about the size of a small fowl, and generally of a dark ashy or sooty colour, and they have remarkably large and strong feet and long claws. They are allied to the "Maleo" of Celebes, of which an account has already been given, but they differ in habits, most of these birds frequenting the scrubby jungles along the sea-shore, where the soil is sandy, and there is a considerable quantity of *débris*, consisting of sticks, shells, seaweed, leaves, &c. Of this rubbish the Megapodius forms immense mounds, often six or eight feet high and

twenty or thirty feet in diameter, which they are enabled
to do with comparative ease by means of their large feet,
with which they can grasp and throw backwards a quantity
of material. In the centre of this mound, at a depth of
two or three feet, the eggs are deposited, and are hatched
by the gentle heat produced by the fermentation of the
vegetable matter of the mound. When I first saw these
mounds in the island of Lombock, I could hardly believe
that they were made by such small birds, but I afterwards
met with them frequently, and have once or twice come
upon the birds engaged in making them. They run a
few steps backwards, grasping a quantity of loose material
in one foot, and throw it a long way behind them.
When once properly buried the eggs seem to be no more
cared for, the young birds working their way up through
the heap of rubbish, and running off at once into the forest.
They come out of the egg covered with thick downy
feathers, and have no tail, although the wings are fully
developed.

I was so fortunate as to discover a new species (Megapodius wallacei), which inhabits Gilolo, Ternate, and
Bouru. It is the handsomest bird of the genus, being
richly banded with reddish brown on the back and wings ;
and it differs from the other species in its habits. It frequents the forests of the interior, and comes down to the
sea-beach to deposit its eggs, but instead of making a

mound, or scratching a hole to receive them, it burrows into the sand to the depth of about three feet obliquely downwards, and deposits its eggs at the bottom. It then loosely covers up the mouth of the hole, and is said by the natives to obliterate and disguise its own footmarks leading to and from the hole, by making many other tracks and scratches in the neighbourhood. It lays its eggs only at night, and at Bouru a bird was caught early one morning as it was coming out of its hole, in which several eggs were found. All these birds seem to be semi-nocturnal, for their loud wailing cries may be constantly heard late into the night and long before daybreak in the morning. The eggs are all of a rusty red colour, and very large for the size of the bird, being generally three or three and a quarter inches long, by two or two and a quarter wide. They are very good eating, and are much sought after by the natives.

Another large and extraordinary bird is the Cassowary, which inhabits the island of Ceram only. It is a stout and strong bird, standing five or six feet high, and covered with long coarse black hair-like feathers. The head is ornamented with a large horny casque or helmet, and the bare skin of the neck is conspicuous with bright blue and red colours. The wings are quite absent, and are replaced by a group of horny black spines like blunt porcupine quills. These birds wander about the vast mountainous forests that cover the island of Ceram, feeding chiefly on fallen fruits, and on insects or crustacea. The female lays from three to five large and beautifully shagreened green eggs upon a bed of leaves, the male and female sitting upon them alternately for about a month. This bird is the helmeted cassowary (Casuarius galeatus) of naturalists, and was for a long time the only species known. Others have since been discovered in New Guinea, New Britain, and North Australia.

It was in the Moluccas that I first discovered undoubted cases of "mimicry" among birds, and these are so curious that I must briefly describe them. It will be as well, however, first to explain what is meant by mimicry in natural history. At page 205 of the first volume of this work, I have described a butterfly which, when at rest, so closely resembles a dead leaf, that it thereby escapes the attacks of its enemies. This is termed a "protective resemblance." If however the butterfly, being itself a savoury morsel to birds, had closely resembled another butterfly which was disagreeable to birds, and therefore never eaten by them, it would be as well protected as if it resembled a leaf; and this is what has been happily termed "mimicry" by Mr. Bates, who first discovered the object of these curious external imitations of one insect by another belonging to a distinct genus or family, and sometimes even to a distinct order. The clear-winged moths which resemble wasps and hornets are the best examples of " mimicry " in our own country.

For a long time all the known cases of exact resemblance of one creature to quite a different one were confined to insects, and it was therefore with great pleasure that I discovered in the island of Bouru two birds which I constantly mistook for each other, and which yet belonged to two distinct and somewhat distant families. One of these is a honeysucker named Tropidorhynchus bouruensis, and the other a kind of oriole, which has been called Mimeta bouruensis. The oriole resembles the honeysucker in the following particulars : the upper and under surfaces of the two birds are exactly of the same tints of dark and light brown ; the Tropidorhynchus has a large bare black patch round the eyes ; this is copied in the Mimeta by a patch of black feathers. The top of the head of the Tropidorhynchus has a scaly appearance from the narrow scale-formed feathers, which are imitated by the broader feathers of the Mimeta having a dusky line down each. The Tropidorhynchus has a pale ruff formed of curious recurved feathers on the nape (which has given the whole genus the name of Friar birds) ; this is represented in the Mimeta by a pale band in the same position. Lastly, the bill of the Tropidorhynchus is raised into a protuberant keel at the base, and the Mimeta has the same character, although it is not a common one in the genus. The result is, that on a

superficial examination the birds are identical, although
they have important structural differences, and cannot be
placed near each other in any natural arrangement.

In the adjacent island of Ceram we find very distinct
species of both these genera, and, strange to say, these
resemble each other quite as closely as do those of Bouru.
The Tropidorhynchus subcornutus is of an earthy brown
colour, washed with ochreish yellow, with bare orbits, dusky
cheeks, and the usual recurved nape-ruff. The Mimeta
forsteni which accompanies it, is absolutely identical in the
tints of every part of the body, and the details are copied
just as minutely as in the former species.

We have two kinds of evidence to tell us which bird in
this case is the model, and which the copy. The honeysuckers are coloured in a manner which is very general in
the whole family to which they belong, while the orioles
seem to have departed from the gay yellow tints so
common among their allies. We should therefore conclude that it is the latter who mimic the former. If
so, however, they must derive some advantage from the
imitation, and as they are certainly weak birds, with small
feet and claws, they may require it. Now the Tropidorhynchi are very strong and active birds, having powerful
grasping claws, and long, curved, sharp beaks. They
assemble together in groups and small flocks, and they have
a very loud bawling note which can be heard at a great

distance, and serves to collect a number together in time of danger.   They are very plentiful and very pugnacious, frequently driving away crows and even hawks, which perch on a tree where a few of them are assembled.   It is very probable, therefore, that the smaller birds of prey have learnt to respect these birds and leave them alone, and it may thus be a great advantage for the weaker and less courageous Mimetas to be mistaken for them.   This being the case, the laws of Variation and Survival of the Fittest, will suffice to explain how the resemblance has been brought about, without supposing any voluntary action on the part of the birds themselves; and those who have read Mr. Darwin's "Origin of Species" will have no difficulty in comprehending the whole process.

The insects of the Moluccas are pre-eminently beautiful, even when compared with the varied and beautiful productions of other parts of the Archipelago.   The grand bird-winged butterflies (Ornithoptera) here reach their maximum of size and beauty, and many of the Papilios, Pieridæ, Danaidæ, and Nymphalidæ are equally pre-eminent.   There is, perhaps, no island in the world so small as Amboyna where so many grand insects are to be found.   Here are three of the very finest Ornithopteræ—priamus, helena, and remus ; three of the handsomest and largest Papilios—ulysses, deiphobus, and gambrisius ; one of the handsomest Pieridæ, Iphias leucippe; the largest of the Danaidæ, Hestia idea; and two unusually large and handsome Nymphalidæ—Diadema pandarus, and Charaxes euryalus. Among its beetles are the extraordinary Euchirus longimanus, whose enormous legs spread over a space of eight inches, and an unusual number of large and handsome Longicorns, Anthribidæ, and Buprestidæ.

The beetles figured on the plate as characteristic of the Moluccas are : 1. A small specimen of the Euchirus longimanus, or Long-armed Chafer, which has been already mentioned in the account of my residence at Amboyna (Chapter XX.). The female has the fore legs of moderate length. 2. A fine weevil, (an undescribed species of Eu-. pholus,) of rich blue and emerald green colours, banded with black. It is a native of Ceram and Goram, and is found on foliage. 3. A female of Xenocerus semiluc-tuosus, one of the Anthribidæ of delicate silky white and black colours. It is abundant on fallen trunks and stumps in Ceram and Amboyna. 4. An unde-scribed species of Xenocerus ; a male, with very long and curious antennæ, and elegant black and white markings. It is found on fallen trunks in Batchian. 5. An un-described species of Arachnobas, a curious genus of weevils peculiar to the Moluccas and New Guinea, and remarkable for their long legs, and their habit of often sitting on leaves, and turning rapidly round the edge to the under-surface when disturbed. It was found in

Eupholus (new species).
Euchirus longimanus, male.

Arachnobas
new species).

Xenocerus semiluctuosus, fem.
Xenocerus new species), male.

MOLUCCAN BEETLES.

Gilolo. All these insects are represented of the natural size.

Like the birds, the insects of the Moluccas show a decided affinity with those of New Guinea rather than with the productions of the great western islands of the Archipelago, but the difference in form and structure between the productions of the east and west is not nearly so marked here as in birds. This is probably due to the more immediate dependence of insects on climate and vegetation, and the greater facilities for their distribution in the varied stages of egg, pupa, and perfect insect. This has led to a general uniformity in the insect-life of the whole Archipelago, in accordance with the general uniformity of its climate and vegetation; while on the other hand the great susceptibility of the insect organization to the action of external conditions has led to infinite detailed modifications of form and colour, which have in many cases given a considerable diversity to the productions of adjacent islands.

Owing to the great preponderance among the birds, of parrots, pigeons, kingfishers, and sunbirds, almost all of gay or delicate colours, and many adorned with the most gorgeous plumage, and to the numbers of very large and showy butterflies which are almost everywhere to be met with, the forests of the Moluccas offer to the naturalist a very striking example of the luxuriance and beauty of

animal life in the tropics. Yet the almost entire absence
of Mammalia, and of such wide-spread groups of birds as
woodpeckers, thrushes, jays, tits, and pheasants, must
convince him that he is in a part of the world which has
in reality but little in common with the great Asiatic
continent, although an unbroken chain of islands seems to
link them to it.

CHAPTER XXVIII.

MACASSAR TO THE ARU ISLANDS IN A NATIVE PRAU

(DECEMBER, 1856.)

IT was the beginning of December, and the rainy season at Macassar had just set in. For nearly three months I had beheld the sun rise daily above the palm-groves, mount to the zenith, and descend like a globe of fire into the ocean, unobscured for a single moment of his course. Now dark leaden clouds had gathered over the whole heavens, and seemed to have rendered him permanently invisible. The strong east winds, warm and dry and dust-laden, which had hitherto blown as certainly as the sun had risen, were now replaced by variable gusty breezes and heavy rains, often continuous for three days and nights together; and the parched and fissured rice stubbles which during the dry weather had extended in every direction for miles around the town, were already so flooded as to be only passable by boats, or by means of a labyrinth of paths on the top of the narrow banks which divided the separate properties.

Five months of this kind of weather might be expected in Southern Celebes, and I therefore determined to seek some more favourable climate for collecting in during that period, and to return in the next dry season to complete my exploration of the district.   Fortunately for me I was in one of the great emporiums of the native trade of the Archipelago.   Rattans from Borneo, sandal-wood and bees'-wax from Flores and Timor, tripang from the Gulf of Carpentaria, cajuputi-oil from Bouru, wild nutmegs and mussoi-bark from New Guinea, are all to be found in the stores of the Chinese and Bugis merchants of Macassar, along with the rice and coffee which are the chief products of the surrounding country.   More important than all these however is the trade to Aru, a group of islands situated on the south-west coast of New Guinea, and of which almost the whole produce comes to Macassar in native vessels. These islands are quite out of the track of all European trade, and are inhabited only by black mop-headed savages, who yet contribute to the luxurious tastes of the most civilized races.   Pearls, mother-of-pearl, and tortoiseshell, find their way to Europe, while edible birds' nests and "tripang" or sea-slug are obtained by shiploads for the gastronomic enjoyment of the Chinese.

The trade to these islands has existed from very early times, and it is from them that Birds of Paradise, of the two kinds known to Linnæus, were first brought.   The

native vessels can only make the voyage once a year,
owing to the monsoons. They leave Macassar in December or January at the beginning of the west monsoon, and
return in July or August with the full strength of the
east monsoon. Even by the Macassar people themselves,
the voyage to the Aru Islands is looked upon as a rather
wild and romantic expedition, full of novel sights and
strange adventures. He who has made it is looked up to
as an authority, and it remains with many the unachieved
ambition of their lives. I myself had hoped rather than
expected ever to reach this " Ultima Thule " of the East ;
and when I found that I really could do so now, had I but
courage to trust myself for a thousand miles' voyage in a
Bugis prau, and for six or seven months among lawless
traders and ferocious savages,—I felt somewhat as I did
when, a schoolboy, I was for the first time allowed to
travel outside the stage-coach, to visit that scene of all that
is strange and new and wonderful to young imaginations
—London !

By the help of some kind friends I was introduced to
the owner of one of the large praus which was to sail in a
few days. He was a Javanese half-caste, intelligent, mild,
and gentlemanly in his manners, and had a young and
pretty Dutch wife, whom he was going to leave behind
during his absence. When we talked about passage money
he would fix no sum, but insisted on leaving it entirely to

me to pay on my return exactly what I liked. "And then," said he, "whether you give me one dollar or a hundred, I shall be satisfied, and shall ask no more."

The remainder of my stay was fully occupied in laying in stores, engaging servants, and making every other preparation for an absence of seven months from even the outskirts of civilization. On the morning of December 13th, when we went on board at daybreak, it was raining hard. We set sail and it came on to blow. Our boat was lost astern, our sails damaged, and the evening found us back again in Macassar harbour. We remained there four days longer, owing to its raining all the time, thus rendering it impossible to dry and repair the huge mat sails. All these dreary days I remained on board, and during the rare intervals when it didn't rain, made myself acquainted with our outlandish craft, some of the peculiarities of which I will now endeavour to describe.

It was a vessel of about seventy tons burthen, and shaped something like a Chinese junk. The deck sloped considerably downward to the bows, which are thus the lowest part of the ship. There were two large rudders, but instead of being placed astern they were hung on the quarters from strong cross beams, which projected out two or three feet on each side, and to which extent the deck overhung the sides of the vessel amidships. The rudders were not hinged but hung with slings of rattan, the friction

of which keeps them in any position in which they are
placed, and thus perhaps facilitates steering. The tillers
were not on deck, but entered the vessel through two
square openings into a lower or half deck about three feet
high, in which sit the two steersmen. In the after part of
the vessel was a low poop, about three and a half feet high,
which forms the captain's cabin, its furniture consisting of
boxes, mats, and pillows. In front of the poop and main-
mast was a little thatched house on deck, about four feet
high to the ridge; and one compartment of this, forming a
cabin six and a half feet long by five and a half wide, I
had all to myself, and it was the snuggest and most comfortable little place I ever enjoyed at sea. It was entered
by a low sliding door of thatch on one side, and had a very
small window on the other. The floor was of split bamboo,
pleasantly elastic, raised six inches above the deck, so as
to be quite dry. It was covered with fine cane mats, for
the manufacture of which Macassar is celebrated; against
the further wall were arranged my gun-case, insect-boxes,
clothes, and books; my mattress occupied the middle, and
next the door were my canteen, lamp, and little store of
luxuries for the voyage; while guns, revolver, and hunting
knife hung conveniently from the roof. During these four
miserable days I was quite jolly in this little snuggery—
more so than I should have been if confined the same time
to the gilded and uncomfortable saloon of a first-class

steamer. Then, how comparatively sweet was everything on board—no paint, no tar, no new rope, (vilest of smells to the qualmish!) no grease, or oil, or varnish ; but instead of these, bamboo and rattan, and coir rope and palm thatch; pure vegetable fibres, which smell pleasantly if they smell at all, and recall quiet scenes in the green and shady forest.

Our ship had two masts, if masts they can be called, which were great moveable triangles. If in an ordinary ship you replace the shrouds and backstay by strong timbers, and take away the mast altogether, you have the arrangement adopted on board a prau. Above my cabin, and resting on cross-beams attached to the masts, was a wilderness of yards and spars, mostly formed of bamboo The mainyard, an immense affair nearly a hundred feet long, was formed of many pieces of wood and bamboo bound together with rattans in an ingenious manner. The sail carried by this was of an oblong shape, and was hung out of the centre, so that when the short end was hauled down on deck the long end mounted high in the air, making up for the lowness of the mast itself. The fore-sail was of the same shape, but smaller. Both these were of matting, and, with two jibs and a fore and aft sail astern of cotton canvas, completed our rig.

The crew consisted of about thirty men, natives of Macassar and the adjacent coasts and islands. They were mostly young, and were short, broad-faced, good-humoured looking fellows. Their dress consisted generally of a pair of trousers only, when at work, and a handkerchief twisted round the head, to which in the evening they would add a thin cotton jacket. Four of the elder men were "jurumudis," or steersmen, who had to squat (two at a time) in the little steerage before described, changing every six hours. Then there was an old man, the "juragan," or captain, but who was really what we should call the first mate; he occupied the other half of the little house on deck. There were about ten respectable men, Chinese or Bugis, whom our owner used to call "his own people." He treated them very well, shared his meals with them, and spoke to them always with perfect politeness; yet they were most of them a kind of slave debtors, bound over by the police magistrate to work for him at mere nominal wages for a term of years till their debts were liquidated. This is a Dutch institution in this part of the world, and seems to work well. It is a great boon to traders, who can do nothing in these thinly-populated regions without trusting goods to agents and petty dealers, who frequently squander them away in gambling and debauchery. The lower classes are almost all in a chronic state of debt. The merchant trusts them again and again, till the amount is something serious, when he brings them to court and has their services allotted to him for its liquidation. The

M 2

debtors seem to think this no disgrace, but rather enjoy their freedom from responsibility, and the dignity of their position under a wealthy and well-known merchant. They trade a little on their own account, and both parties seem to get on very well together. The plan seems a more sensible one than that which we adopt, of effectually preventing a man from earning anything towards paying his debts by shutting him up in a jail.

My own servants were three in number. Ali, the Malay boy whom I had picked up in Borneo, was my head man. He had already been with me a year, could turn his hand to anything, and was quite attentive and trustworthy. He was a good shot, and fond of shooting, and I had taught him to skin birds very well. The second, named Baderoon, was a Macassar lad, also a pretty good boy, but a desperate gambler. Under pretence of buying a house for his mother, and clothes for himself, he had received four months' wages about a week before we sailed, and in a day or two gambled away every dollar of it. He had come on board with no clothes, no betel, or tobacco, or salt fish, all which necessary articles I was obliged to send Ali to buy for him. These two lads were about sixteen, I should suppose; the third was younger, a sharp little rascal named Baso, who had been with me a month or two, and had learnt to cook tolerably. He was to fulfil the important office of cook and housekeeper, for I could not get any regular servants to go to such a terribly remote country; one might as well ask a *chef de cuisine* to go to Patagonia.

On the fifth day that I had spent on board (Dec. 15th) the rain ceased, and final preparations were made for starting. Sails were dried and furled, boats were constantly coming and going, and stores for the voyage, fruit, vegetables, fish, and palm sugar, were taken on board. In the afternoon two women arrived with a large party of friends and relations, and at parting there was a general nose-rubbing (the Malay kiss), and some tears shed. These were promising symptoms for our getting off the next day; and accordingly, at three in the morning, the owner came on board, the anchor was immediately weighed, and by four we set sail. Just as we were fairly off and clear of the other praus, the old juragan repeated some prayers, all around responding with "Allah il Allah," and a few strokes on a gong as an accompaniment, concluding with all wishing each other "Salaamat jalan," a safe and happy journey. We had a light breeze, a calm sea, and a fine morning, a prosperous commencement of our voyage of about a thousand miles to the far-famed Aru Islands.

The wind continued light and variable all day, with a calm in the evening before the land breeze sprang up. We were then passing the island of "Tanakaki" (foot of the land), at the extreme south of this part of Celebes. There are some dangerous rocks here, and as I was standing by the bulwarks, I happened to spit over the side; one of the men begged I would not do so just now, but spit on deck, as they were much afraid of this place. Not quite comprehending, I made him repeat his request, when, seeing he was in earnest, I said, "Very well, I suppose there are 'hantus' (spirits) here." "Yes," said he, "and they don't like anything to be thrown overboard; many a prau has been lost by doing it." Upon which I promised to be very careful. At sunset the good Mahometans on board all repeated a few words of prayer with a general chorus, reminding me of the pleasing and impressive "Ave Maria" of Catholic countries.

Dec. 20*th.*—At sunrise we were opposite the Bontyne mountain, said to be one of the highest in Celebes. In the afternoon we passed the Salayer Straits and had a little squall, which obliged us to lower our huge mast, sails, and heavy yards. The rest of the evening we had a fine west wind, which carried us on at near five knots an hour, as much as our lumbering old tub can possibly go.

Dec. 21*st.*—A heavy swell from the south-west rolling us about most uncomfortably. A steady wind was blowing, however, and we got on very well.

Dec. 22*d.*—The swell had gone down. We passed Boutong, a large island, high, woody, and populous, the native place of some of our crew. A small prau returning from Bali to the island of Goram overtook us. The

DOBBO, IN THE TRADING SEASON.

nakoda (captain) was known to our owner. They had been two years away, but were full of people, with several black Papuans on board. At 6 P.M. we passed Wangi-wangi, low but not flat, inhabited and subject to Boutong. We had now fairly entered the Molucca Sea. After dark it was a beautiful sight to look down on our rudders, from which rushed eddying streams of phosphoric light gemmed with whirling sparks of fire. It resembled (more nearly than anything else to which I can compare it) one of the large irregular nebulous star-clusters seen through a good tele-scope, with the additional attraction of ever-changing form and dancing motion.

Dec. 23d.—Fine red sunrise; the island we left last evening barely visible behind us. The Goram prau about a mile south of us. They have no compass, yet they have kept a very true course during the night. Our owner tells me they do it by the swell of the sea, the direction of which they notice at sunset, and sail by it during the night. In these seas they are never (in fine weather) more than two days without seeing land. Of course adverse winds or currents sometimes carry them away, but they soon fall in with some island, and there are always some old sailors on board who know it, and thence take a new course. Last night a shark about five feet long was caught, and this morning it was cut up and cooked. In the afternoon they got another, and I had a little fried, and found it firm and dry, but very palatable. In the evening the sun set in a heavy bank of clouds, which, as darkness came on, assumed a fearfully black appearance. According to custom, when strong wind or rain is expected, our large sails were furled, and with their yards let down on deck, and a small square foresail alone kept up. The great mat sails are most awkward things to manage in rough weather. The yards which support them are seventy feet long, and of course very heavy ; and the only way to furl them being to roll up the sail on the boom, it is a very dangerous thing to have them standing when overtaken by a squall. Our crew, though numerous enough for a vessel of 700 instead of one of 70 tons, have it very much their own way, and there seems to be seldom more than a dozen at work at a time. When anything important is to be done, however, all start up willingly enough, but then all think themselves at liberty to give their opinion, and half a dozen voices are heard giving orders, and there is such a shrieking and confusion that it seems wonderful anything gets done at all.

Considering we have fifty men of several tribes and tongues on board, wild, half-savage looking fellows, and few of them feeling any of the restraints of morality or education, we get on wonderfully well. There is no fighting or quarrelling, as there would certainly be among the same number of Europeans with as little restraint upon their actions, and there is scarcely any of that noise and excitement which might be expected. In fine weather the greater part of them are quietly enjoying themselves— some are sleeping under the shadow of the sails ; others, in little groups of three or four, are talking or chewing betel ; one is making a new handle to his chopping-knife, another is stitching away at a new pair of trousers or a shirt, and all are as quiet and well-conducted as on board the best-ordered English merchantman. Two or three take it by turns to watch in the bows and see after the braces and halyards of the great sails ; the two steersmen are below in the steerage ; our captain, or the juragan, gives the course, guided partly by the compass and partly by the direction of the wind, and a watch of two or three on the poop look after the trimming of the sails and call out the hours by the water-clock. This is a very ingenious contrivance, which measures time well in both rough weather and fine. It is simply a bucket half filled with water, in which floats the half of a well-scraped cocoa-nut shell. In the bottom of this shell is a very small hole, so that when placed to float in the bucket a fine thread of water squirts up into it. This gradually fills the shell, and the size of the hole is so adjusted to the capacity of the vessel that, exactly at the end of an hour, plump it goes to the bottom. The watch then cries out the number of hours from sunrise, and sets the shell afloat again empty. This is a very good measurer of time. I tested it with my watch and found that it hardly varied a minute from one hour to another, nor did the motion of the vessel have any effect upon it, as the water in the bucket of course kept level. It has a great advantage for a rude people in being easily understood, in being rather bulky and easy to see, and in the final submergence being accompanied with a little bubbling and commotion of the water, which calls the attention to it. It is also quickly replaced if lost while in harbour.

Our captain and owner I find to be a quiet, good-tempered man, who seems to get on very well with all about him. When at sea he drinks no wine or spirits, but indulges only in coffee and cakes, morning and after-noon, in company with his supercargo and assistants. He is a man of some little education, can read and write well both Dutch and Malay, uses a compass, and has a chart. He has been a trader to Aru for many years, and is well known to both Europeans and natives in this part of the world.

Dec. 24th.—Fine, and little wind. No land in sight for the first time since we left Macassar. At noon calm, with heavy showers, in which our crew wash their clothes, and in the afternoon the prau is covered with shirts, trousers, and sarongs of various gay colours. I made a discovery to-day which at first rather alarmed me. The two ports,

or openings, through which the tillers enter from the
lateral rudders are not more than three or four feet above
the surface of the water, which thus has a free entrance
into the vessel. I of course had imagined that this open
space from one side to the other was separated from the
hold by a water-tight bulkhead, so that a sea entering
might wash out at the further side, and do no more harm
than give the steersmen a drenching. To my surprise
and dismay, however, I find that it is completely open to
the hold, so that half-a-dozen seas rolling in on a stormy
night would nearly, or quite, swamp us. Think of a vessel
going to sea for a month with two holes, each a yard
square, into the hold, at three feet above the water-line,—
holes, too, which cannot possibly be closed. But our
captain says all praus are so; and though he acknowledges
the danger, " he does not know how to alter it—the people
are used to it; he does not understand praus so well as
they do, and if such a great alteration were made, he
should be sure to have difficulty in getting a crew!" This
proves at all events that praus must be good sea-boats,
for the captain has been continually making voyages in
them for the last ten years, and says he has never known
water enough enter to do any harm.

Dec. 25th.—Christmas-day dawned upon us with gusts
of wind, driving rain, thunder and lightning, added to
which a short confused sea made our queer vessel pitch

and roll very uncomfortably. About nine o'clock, however, it cleared up, and we then saw ahead of us the fine island of Bouru, perhaps forty or fifty miles distant, its mountains wreathed with clouds, while its lower lands were still invisible. The afternoon was fine, and the wind got round again to the west; but although this is really the west monsoon, there is no regularity or steadiness about it, calms and breezes from every point of the compass continually occurring. The captain, though nominally a Protestant, seemed to have no idea of Christmas-day as a festival. Our dinner was of rice and curry as usual, and an extra glass of wine was all I could do to celebrate it.

Dec. 26th.—Fine view of the mountains of Bouru, which we have now approached considerably. Our crew seem rather a clumsy lot. They do not walk the deck with the easy swing of English sailors, but hesitate and stagger like landsmen. In the night the lower boom of our mainsail broke, and they were all the morning repairing it. It consisted of two bamboos lashed together, thick end to thin, and was about seventy feet long. The rigging and arrangement of these praus contrasts strangely with that of European vessels, in which the various ropes and spars, though much more numerous, are placed so as not to interfere with each other's action. Here the case is quite different; for though there are no shrouds or stays to complicate the matter, yet scarcely anything can be done without first clearing something else out of the way. The large sails cannot be shifted round to go on the other tack without first hauling down the jibs, and the booms of the fore and aft sails have to be lowered and completely detached to perform the same operation. Then there are always a lot of ropes foul of each other, and all the sails can never be set (though they are so few) without a good part of their surface having the wind kept out of them by others. Yet praus are much liked even by those who have had European vessels, because of their cheapness both in first cost and in keeping up; almost all repairs can be done by the crew, and very few European stores are required.

Dec. 28th.—This day we saw the Banda group, the volcano first appearing,—a perfect cone, having very much the outline of the Egyptian pyramids, and looking almost as regular. In the evening the smoke rested over its summit like a small stationary cloud. This was my first view of an active volcano, but pictures and panoramas have so impressed such things on one's mind, that when we at length behold them they seem nothing extraordinary.

Dec. 30th.—Passed the island of Teor, and a group near it, which are very incorrectly marked on the charts. Flying-fish were numerous to-day. It is a smaller species than that of the Atlantic, and more active and elegant in its motions.   As they skim along the surface they turn on their sides, so as fully to display their beautiful fins, taking a flight of about a hundred yards, rising and falling in a most graceful manner.   At a little distance they exactly resemble swallows, and no one who sees them can doubt that they really do fly, not merely descend in an oblique direction from the height they gain by their first spring. In the evening an aquatic bird, a species of booby (Sula fiber.) rested on our hen-coop, and was caught by the neck by one of my boys.

Dec. 31*st*.—At daybreak the Ké Islands (pronounced kay) were in sight, where we are to stay a few days. About noon we rounded the northern point, and endeavoured to coast along to the anchorage; but being now on the leeward side of the island, the wind came in violent irregular gusts, and then leaving us altogether, we were carried back by a strong current.   Just then two boats-load of natives appeared, and our owner having agreed with them to tow us into harbour, they tried to do so, assisted by our own boat, but could make no way.   We were therefore obliged to anchor in a very dangerous place on a rocky bottom, and we were engaged till nearly dark getting hawsers secured to some rocks under water.   The coast of Ké along which we had passed was very picturesque.   Light coloured limestone rocks rose abruptly from the water to the height of several hundred feet, every-

where broken into jutting peaks and pinnacles, weatherworn into sharp points and honeycombed surfaces, and
clothed throughout with a most varied and luxuriant
vegetation. The cliffs above the sea offered to our view
screw-pines and arborescent Liliaceæ of strange forms,
mingled with shrubs and creepers; while the higher
slopes supported a dense growth of forest trees. Here and
there little bays and inlets presented beaches of dazzling
whiteness. The water was transparent as crystal, and tinged
the rock-strewn slope which plunged steeply into its
unfathomable depths with colours varying from emerald
to lapis-lazuli. The sea was calm as a lake, and the
glorious sun of the tropics threw a flood of golden light
over all. The scene was to me inexpressibly delightful.
I was in a new world, and could dream of the wonderful
productions hid in those rocky forests, and in those azure
abysses. But few European feet had ever trodden the
shores I gazed upon; its plants, and animals, and men
were alike almost unknown, and I could not help speculating on what my wanderings there for a few days might
bring to light.

CHAPTER XXIX.

THE KÉ ISLANDS.

(JANUARY 1857.)

THE native boats that had come to meet us were three or four in number, containing in all about fifty men. They were long canoes, with the bow and stern rising up into a beak six or eight feet high, decorated with shells and waving plumes of cassowaries hair. I now had my first view of Papuans in their own country, and in less than five minutes was convinced that the opinion already arrived at by the examination of a few Timor and New Guinea slaves was substantially correct, and that the people I now had an opportunity of comparing side by side belonged to two of the most distinct and strongly marked races that the earth contains. Had I been blind, I could have been certain that these islanders were not Malays. The loud, rapid, eager tones, the incessant motion, the intense vital activity manifested in speech and action, are the very antipodes of the quiet, unimpulsive, unanimated Malay. These Ké men came up singing and shouting, dipping their paddles deep in the water and throwing up clouds of spray ; as they approached nearer they stood up in their canoes and increased their noise and gesticulations; and on coming alongside, without asking leave, and without a moment's hesitation, the greater part of them scrambled up on our deck just as if they were come to take possession of a captured vessel. Then commenced a scene of indescribable confusion. These forty black, naked, mop-headed savages seemed intoxicated with joy and excitement. Not one of them could remain still for a moment. Every individual of our crew was in turn surrounded and examined, asked for tobacco or arrack, grinned at, and deserted for another. All talked at once, and our captain was regularly mobbed by the chief men, who wanted to be employed to tow us in, and who begged vociferously to be paid in advance. A few presents of tobacco made their eyes glisten; they would express their satisfaction by grins and shouts, by rolling on deck, or by a headlong leap overboard. School-boys on an unexpected holiday, Irishmen at a fair, or midshipmen on shore, would give but a faint idea of the exuberant animal enjoyment of these people.

Under similar circumstances Malays *could* not behave as these Papuans did. If they came on board a vessel (after asking permission), not a word would be at first spoken, except a few compliments, and only after some time, and very cautiously, would any approach be made to business.

One would speak at a time, with a low voice and great
deliberation, and the mode of making a bargain would be
by quietly refusing all your offers, or even going away
without saying another word about the matter, unless you
advanced your price to what they were willing to accept.
Our crew, many of whom had not made the voyage before,
seemed quite scandalized at such unprecedented bad
manners, and only very gradually made any approach to
fraternization with the black fellows. They reminded me
of a party of demure and well-behaved children suddenly
broken in upon by a lot of wild romping, riotous boys,
whose conduct seems most extraordinary and very naughty!

These moral features are more striking and more conclusive of absolute diversity than even the physical
contrast presented by the two races, though that is sufficiently remarkable. The sooty blackness of the skin, the
mop-like head of frizzly hair, and, most important of all,
the marked form of countenance of quite a different type
from that of the Malay, are what we cannot believe to
result from mere climatal or other modifying influences on
one and the same race. The Malay face is of the Mongolian type, broad and somewhat flat. The brows are
depressed, the mouth wide, but not projecting, and the nose
small and well formed but for the great dilatation of the
nostrils. The face is smooth, and rarely develops the trace
of a beard; the hair black, coarse, and perfectly straight.

The Papuan, on the other hand, has a face which we may say is compressed and projecting. The brows are protuberant and overhanging, the mouth large and prominent while the nose is very large, the apex elongated downwards, the ridge thick, and the nostrils large. It is an obtrusive and remarkable feature in the countenance, the very reverse of what obtains in the Malay face. The twisted beard and frizzly hair complete this remarkable contrast. Here then I had reached a new world, inhabited by a strange people. Between the Malayan tribes, among whom I had for some years been living, and the Papuan races, whose country I had now entered, we may fairly say that there is as much difference, both moral and physical, as between the red Indians of South America and the negroes of Guinea on the opposite side of the Atlantic.

Jan. 1st, 1857.—This has been a day of thorough enjoyment. I have wandered in the forests of an island rarely seen by Europeans. Before daybreak we left our anchorage, and in an hour reached the village of Har, where we were to stay three or four days. The range of hills here receded so as to form a small bay, and they were broken up into peaks and hummocks with intervening flats and hollows. A broad beach of the whitest sand lined the inner part of the bay, backed by a mass of cocoa-nut palms, among which the huts were concealed, and surmounted by a dense and varied growth of timber. Canoes and boats of various sizes were drawn up on the beach, and one or two idlers, with a few children and a dog, gazed at our prau as we came to an anchor.

When we went on shore the first thing that attracted us was a large and well-constructed shed, under which a long boat was being built, while others in various stages of completion were placed at intervals along the beach. Our captain, who wanted two of moderate size for the trade among the islands at Aru, immediately began bargaining for them, and in a short time had arranged the number of brass guns, gongs, sarongs, handkerchiefs, axes, white plates, tobacco, and arrack, which he was to give for a pair which could be got ready in four days. We then went to the village, which consisted only of three or four huts, situated immediately above the beach on an irregular rocky piece of ground overshadowed with cocoa-nuts, palms, bananas, and other fruit trees. The houses were very rude, black, and half rotten, raised a few feet on posts with low sides of bamboo or planks, and high thatched roofs. They had small doors and no windows, an opening under the projecting gables letting the smoke out and a little light in. The floors were of strips of bamboo, thin, slippery, and elastic, and so weak that my feet were in danger of plunging through at every step. Native boxes of pandanus-leaves and slabs of palm pith, very neatly constructed, mats of the same, jars and cooking pots of native pottery, and a few European plates and basins, were the whole furniture, and the interior was throughout dark and smoke-blackened, and dismal in the extreme.

Accompanied by Ali and Baderoon, I now attempted to make some explorations, and we were followed by a train of boys eager to see what we were going to do. The most trodden path from the beach led us into a shady hollow, where the trees were of immense height and the under-growth scanty. From the summits of these trees came at intervals a deep booming sound, which at first puzzled us, but which we soon found to proceed from some large pigeons. My boys shot at them, and after one or two misses, brought one down. It was a magnificent bird twenty inches long, of a bluish white colour, with the back wings and tail intense metallic green, with golden, blue, and violet reflexions, the feet coral red, and the eyes golden yellow. It is a rare species, which I have named Carpophaga concinna, and is found only in a few small islands, where, however, it abounds. It is the same species which in the island of Banda is called the nutmeg-pigeon, from its habit of devouring the fruits, the seed or nutmeg being thrown up entire and uninjured. Though these pigeons have a narrow beak, yet their jaws and throat are so extensible that they can swallow fruits of very large size. I had before shot a species much smaller than this one, which had a number of hard globular palm-fruits in its crop, each more than an inch in diameter.

A little further the path divided into two, one leading along the beach, and across mangrove and sago swamps, the other rising to cultivated grounds. We therefore returned. and taking a fresh departure from the village, endeavoured to ascend the hills and penetrate into the interior. The path, however, was a most trying one. Where there was earth, it was a deposit of reddish clay overlying the rock, and was worn so smooth by the attrition of naked feet that my shoes could obtain no hold on the sloping surface. A little farther we came to the bare rock, and this was worse, for it was so rugged and broken, and so honeycombed and weatherworn into sharp points and angles, that my boys, who had gone barefooted all their lives, could not stand it. Their feet began to bleed, and I saw that if I did not want them completely lamed it would be wise to turn back. My own shoes, which were rather thin, were but a poor protection, and would soon have been cut to pieces; yet our little naked guides tripped along with the greatest ease and unconcern, and seemed much astonished at our effeminacy in not being able to take a walk which to them was a perfectly agreeable one. During the rest of our stay in the island we were obliged to confine ourselves to the vicinity of the shore and the cultivated grounds, and those more level portions of the forest where a little soil had accumulated and the rock had been less exposed to atmospheric action.

The island of Ké (pronounced exactly as the letter K, but erroneously spelt in our maps Key or Ki) is long and narrow, running in a north and south direction, and consists almost entirely of rock and mountain. It is everywhere covered with luxuriant forests, and in its bays and inlets the sand is of dazzling whiteness, resulting from the decomposition of the coralline limestone of which it is entirely composed. In all the little swampy inlets and valleys sago trees abound, and these supply the main subsistence of the natives, who grow no rice, and have scarcely any other cultivated products but cocoa-nuts, plantains, and yams. From the cocoa-nuts, which surround every hut, and which thrive exceedingly on the porous limestone soil and under the influence of salt breezes, oil is made which is sold at a good price to the Aru traders, who all touch here to lay in their stock of this article, as well as to purchase boats and native crockery. Wooden bowls, pans, and trays are also largely made here, hewn out of solid blocks of wood with knife and adze; and these are carried to all parts of the Moluccas. But the art in which the natives of Ké pre-eminently excel is that of boat-building. Their forests supply abundance of fine timber, though probably not more so than many other islands, and from some unknown causes these remote savages have come to excel in what seems a very difficult art. Their small canoes are beautifully formed, broad and low in the centre, but rising at each end, where they terminate in high-pointed beaks more or less carved, and ornamented with a plume of feathers. They are not hollowed out of a tree, but are regularly built of planks running from end to end, and so accurately fitted that it is often difficult to find a place where a knife-blade can be inserted between the joints. The larger ones are from 20 to 30 tons burthen, and are finished ready for sea. without a nail or particle of iron being used, and with no other tools than axe, adze, and auger. These vessels are handsome to look at, good sailers, and admirable sea-boats, and will make long voyages with perfect safety, traversing the whole Archipelago from New Guinea to Singapore in seas which, as every one who has sailed much in them can testify, are not so smooth and tempest free as word-painting travellers love to represent them.

The forests of Ké produce magnificent timber, tall, straight, and durable, of various qualities, some of which are said to be superior to the best Indian teak. To make each pair of planks used in the construction of the larger boats an entire tree is consumed. It is felled, often miles away from the shore, cut across to the proper length, and then hewn longitudinally into two equal portions. Each of these forms a plank by cutting down with the axe to a

uniform thickness of three or four inches, leaving at first a
solid block at each end to prevent splitting. Along the
centre of each plank a series of projecting pieces are left,
standing up three or four inches, about the same width, and
a foot long ; these are of great importance in the construction of the vessel. When a sufficient number of planks
have been made, they are laboriously dragged through the
forest by three or four men each to the beach, where the
boat is to be built. A foundation piece, broad in the
middle and rising considerably at each end, is first laid on
blocks and properly shored up. The edges of this are
worked true and smooth with the adze, and a plank, properly curved and tapering at each end, is held firmly up
against it, while a line is struck along it which allows it
to be cut so as to fit exactly. A series of auger holes,
about as large as one's finger, are then bored along the
opposite edges, and pins of very hard wood are fitted to
these, so that the two planks are held firmly, and can be
driven into the closest contact ; and difficult as this seems
to do without any other aid than rude practical skill in
forming each edge to the true corresponding curves, and in
boring the holes so as exactly to match both in position
and direction, yet so well is it done that the best European
shipwright cannot produce sounder or closer-fitting joints.
The boat is built up in this way by fitting plank to
plank till the proper height and width are obtained.

We have now a skin held together entirely by the hard-wood pins connecting the edges of the planks, very strong and elastic, but having nothing but the adhesion of these pins to prevent the planks gaping. In the smaller boats seats, in the larger ones cross-beams, are now fixed. They are sprung into slight notches cut to receive them, and are further secured to the projecting pieces of the plank below by a strong lashing of rattan. Ribs are now formed of single pieces of tough wood chosen and trimmed so as exactly to fit on to the projections from each plank, being slightly notched to receive them, and securely bound to them by rattans passed through a hole in each projecting piece close to the surface of the plank. The ends are closed against the vertical prow and stern posts, and further secured with pegs and rattans, and then the boat is complete; and when fitted with rudders, masts, and thatched covering, is ready to do battle with the waves. A careful consideration of the principle of this mode of construction, and allowing for the strength and binding qualities of rattan (which resembles in these respects wire rather than cordage), makes me believe that a vessel care-fully built in this manner is actually stronger and safer than one fastened in the ordinary way with nails.

During our stay here we were all very busy. Our captain was daily superintending the completion of his two small praus. All day long native boats were coming

with fish, cocoa-nuts, parrots and lories, earthen pans,
sirip leaf, wooden bowls, and trays, &c. &c., which every
one of the fifty inhabitants of our prau seemed to be
buying on his own account, till all available and most
unavailable space of our vessel was occupied with these
miscellaneous articles : for every man on board a prau
considers himself at liberty to trade, and to carry with
him whatever he can afford to buy.

Money is unknown and valueless here—knives, cloth,
and arrack forming the only medium of exchange, with
tobacco for small coin. Every transaction is the subject of
a special bargain, and the cause of much talking. It is
absolutely necessary to offer very little, as the natives are
never satisfied till you add a little more. They are then
far better pleased than if you had given them twice the
amount at first and refused to increase it.

I, too, was doing a little business, having persuaded
some of the natives to collect insects for me; and when
they really found that I gave them most fragrant tobacco
for worthless black and green beetles, I soon had scores of
visitors, men, women, and children, bringing bamboos full
of creeping things, which, alas! too frequently had eaten
each other into fragments during the tedium of a day's
confinement. Of one grand new beetle, glittering with
ruby and emerald tints, I got a large quantity, having first
detected one of its wing-cases ornamenting the outside of

a native's tobacco pouch.   It was quite a new species, and had not been found elsewhere than on this little island. It is one of the Buprestidæ, and has been named Cyphogastra calepyga.

Each morning after an early breakfast I wandered by myself into the forest, where I found delightful occupation in capturing the large and handsome butterflies, which were tolerably abundant, and most of them new to me ; for I was now upon the confines of the Moluccas and New Guinea,—a region the productions of which were, then among the most precious and rare in the cabinets of Europe.   Here my eyes were feasted for the first time with splendid scarlet lories on the wing, as well as by the sight of that most imperial butterfly, the "Priamus" of collectors, or a closely allied species, but flying so high that I did not succeed in capturing a specimen.   One of them was brought me in a bamboo, boxed up with a lot of beetles, and of course torn to pieces.   The principal drawback of the place for a collector is the want of good paths, and the dreadfully rugged character of the surface, requiring the attention to be so continually directed to securing a footing, as to make it very difficult to capture active winged things, who pass out of reach while one is glancing to see that the next step may not plunge one into a chasm or over a precipice.   Another inconvenience is that there are no running streams, the rock being of so

porous a nature that the surface-water everywhere penetrates its fissures ; at least such is the character of the
neighbourhood we visited, the only water being small
springs trickling out close to the sea-beach.

In the forests of Ké, arboreal Liliaceæ and Pandanaceæ
abound, and give a character to the vegetation in the more
exposed rocky places. Flowers were scarce, and there
were not many orchids, but I noticed the fine white
butterfly-orchis, Phalænopsis grandiflora, or a species
closely allied to it. The freshness and vigour of the
vegetation was very pleasing, and on such an arid rocky
surface was a sure indication of a perpetually humid
climate. Tall clean trunks, many of them buttressed, and
immense trees of the fig family, with aërial roots stretching
out and interlacing and matted together for fifty or a
hundred feet above the ground, were the characteristic
features ; and there was an absence of thorny shrubs and
prickly rattans, which would have made these wilds very
pleasant to roam in, had it not been for the sharp honeycombed rocks already alluded to In damp places a fine
undergrowth of broad-leaved herbaceous plants was found,
about which swarmed little green lizards, with tails of the
most " heavenly blue," twisting in and out among the
stalks and foliage so actively that I often caught glimpses
of their tails only, when they startled me by their resemblance to small snakes. Almost the only sounds in these

primæval woods proceeded from two birds, the red lories, who utter shrill screams like most of the parrot tribe, and the large green nutmeg-pigeon, whose voice is either a loud and deep boom, like two notes struck upon a very large gong, or sometimes a harsh toad-like croak, altogether peculiar and remarkable. Only two quadrupeds are said by the natives to inhabit the island—a wild pig and a Cuscus, or Eastern opossum, of neither of which could I obtain specimens.

The insects were more abundant, and very interesting. Of butterflies I caught thirty-five species, most of them new to me, and many quite unknown in European collections. Among them was the fine yellow and black Papilio euchenor, of which but few specimens had been previously captured, and several other handsome butterflies of large size, as well as some beautiful little " blues," and some brilliant day-flying moths. The beetle tribe were less abundant, yet I obtained some very fine and rare species. On the leaves of a slender shrub in an old clearing I found several fine blue and black beetles of the genus Eupholus, which almost rival in beauty the diamond beetles of South America. Some cocoa-nut palms in blossom on the beach were frequented by a fine green floral beetle (Lomaptera papua), which, when the flowers were shaken, flew off like a small swarm of bees. I got one of our crew to climb up the tree, and he brought me a good number in his hand;

and seeing they were valuable, I sent him up again with
my net to shake the flowers into, and thus secured a large
quantity.   My best capture, however, was the superb
insect of the Buprestis family, already mentioned as
having been obtained from the natives, who told me they
found it in rotten trees in the mountains.

In the forest itself the only common and conspicuous
coleoptera were two tiger beetles.   One, Therates labiata,
was much larger than our green tiger beetle, of a purple
black colour, with green metallic glosses, and the broad
upper lip of a bright yellow.   It was always found upon
foliage, generally of broad-leaved herbaceous plants, and in
damp and gloomy situations, taking frequent short flights
from leaf to leaf, and preserving an alert attitude, as if
always looking out for its prey.   Its vicinity could be immediately ascertained, often before it was seen, by a very
pleasant odour, like otto of roses, which it seems to emit
continually, and which may probably be attractive to the
small insects on which it feeds.   The other, Tricondyla
aptera, is one of the most curious forms in the family of
the Cicindelidæ, and is almost exclusively confined to the
Malay islands.   In shape it resembles a very large ant,
more than an inch long, and of a purple black colour.
Like an ant also it is wingless, and is generally found
ascending trees, passing around the trunks in a spiral
direction when approached, to avoid capture, so that it

requires a sudden run and active fingers to secure a
specimen. This species emits the usual fetid odour of the
ground beetles. My collections during our four days' stay
at Ké were as follow:—Birds, 13 species; insects, 194
species ; and 3 kinds of land-shells.

There are two kinds of people inhabiting these islands
—the indigenes, who have the Papuan characters strongly
marked, and who are pagans ; and a mixed race, who are
nominally Mahometans, and wear cotton clothing, while
the former use only a waist cloth of cotton or bark. These
Mahometans are said to have been driven out of Banda by
the early European settlers. They were probably a brown
race, more allied to the Malays, and their mixed descendants here exhibit great variations of colour, hair, and
features, graduating between the Malay and Papuan types.
It is interesting to observe the influence of the early
Portuguese trade with these countries in the words of
their language, which still remain in use even among these
remote and savage islanders. " Lenço " for handkerchief,
and "faca" for knife are here used to the exclusion of the
proper Malay terms. The Portuguese and Spaniards were
truly wonderful conquerors and colonizers. They effected
more rapid changes in the countries they conquered than
any other nations of modern times, resembling the Romans
in their power of impressing their own language, religion,
and manners on rude and barbarous tribes.

The striking contrast of character between these people and the Malays is exemplified in many little traits. One day when I was rambling in the forest, an old man stopped to look at me catching an insect. He stood very quiet till I had pinned and put it away in my collecting box, when he could contain himself no longer, but bent almost double, and enjoyed a hearty roar of laughter. Every one will recognise this as a true negro trait. A Malay would have stared, and asked with a tone of bewilderment what I was doing, for it is but little in his nature to laugh, never heartily, and still less at or in the presence of a stranger, to whom, however, his disdainful glances or whispered remarks are less agreeable than the most boisterous open expression of merriment. The women here were not so much frightened at strangers, or made to keep themselves so much secluded as among the Malay races; the children were more merry and had the "nigger grin," while the noisy confusion of tongues among the men, and their excitement on very ordinary occasions, are altogether removed from the general taciturnity and reserve of the Malay.

The language of the Ke people consists of words of one, two, or three syllables in about equal proportions, and has many aspirated and a few guttural sounds. The different villages have slight differences of dialect, but they are mutually intelligible, and, except in words that have

evidently been introduced during a long-continued commercial intercourse, seem to have no affinity whatever with
the Malay languages.

Jan. 6th.—The small boats being finished, we sailed
for Aru at 4 P.M., and as we left the shores of Ké had a
fine view of its rugged and mountainous character; ranges
of hills, three or four thousand feet high, stretching south-
wards as fai as the eye could reach, everywhere covered
with a lofty, dense, and unbroken forest. We had very
light winds, and it therefore took us thirty hours to make
the passage of sixty miles to the low, or flat, but equally
forest-covered Aru Islands, where we anchored in the
harbour of Dobbo at nine in the evening of the next day.

My first voyage in a prau being thus satisfactorily
terminated, I must, before taking leave of it for some
months, bear testimony to the merits of the queer old-
world vessel. Setting aside all ideas of danger, which is
probably, after all, not more than in any other craft, I
must declare that I have never, either before or since,
made a twenty days' voyage so pleasantly, or perhaps,
more correctly speaking, with so little discomfort. This I
attribute chiefly to having my small cabin on deck, and
entirely to myself, to having my own servants to wait
upon me, and to the absence of all those marine-store
smells of paint, pitch, tallow, and new cordage, which are
to me insupportable. Something is also to be put down

to freedom from all restraint of dress, hours of meals, &c.,
and to the civility and obliging disposition of the captain.
I had agreed to have my meals with him, but whenever I
wished it I had them in my own berth, and at what
hours I felt inclined. The crew were all civil and good-
tempered, and with very little discipline everything went
on smoothly, and the vessel was kept very clean and in
pretty good order, so that on the whole I was much
delighted with the trip, and was inclined to rate the
luxuries of the semi-barbarous prau as surpassing those of
the most magnificent screw-steamer, that highest result
of our civilization.

CHAPTER XXX.

ON the 8th of January, 1857, I landed at Dobbo, the trading settlement of the Bugis and Chinese, who annually visit the Aru Islands. It is situated on the small island of Wamma, upon a spit of sand which projects out to the north, and is just wide enough to contain three rows of houses. Though at first sight a most strange and desolate-looking place to build a village on, it has many advantages. There is a clear entrance from the west among the coral reefs that border the land, and there is good anchorage for vessels, on one side of the village or the other, in both the east and west monsoons Being fully exposed to the sea-breezes in three directions it is healthy, and the soft sandy beach offers great facilities for hauling up the praus, in order to secure them from sea-worms and prepare them for the homeward voyage. At its southern extremity the sand-bank merges in the beach of the island, and is backed by a luxuriant growth

of lofty forest. The houses are of various sizes, but are
all built after one pattern, being merely large thatched
sheds, a small portion of which, next the entrance, is used
as a dwelling, while the rest is parted off, and often
divided by one or two floors, in order better to stow away
merchandise and native produce.

As we had arrived early in the season, most of the
houses were empty, and the place looked desolate in the
extreme—the whole of the inhabitants who received us
on our landing amounting to about half-a-dozen Bugis and
Chinese. Our captain, Herr Warzbergen, had promised
to obtain a house for me, but unforeseen difficulties presented themselves. One which was to let had no roof,
and the owner, who was building it on speculation, could
not promise to finish it in less than a month. Another,
of which the owner was dead, and which I might therefore take undisputed possession of as the first comer,
wanted considerable repairs, and no one could be found
to do the work, although about four times its value was
offered. The captain, therefore, recommended me to take
possession of a pretty good house near his own, whose
owner was not expected for some weeks; and as I was
anxious to be on shore, I immediately had it cleared out,
and by evening had all my things housed, and was
regularly installed as an inhabitant of Dobbo. I had
brought with me a cane chair, and a few light boards,

which were soon rigged up into a table and shelves. A
broad bamboo bench served as sofa and bedstead, my
boxes were conveniently arranged, my mats spread on the
floor, a window cut in the palm-leaf wall to light my
table, and though the place was as miserable and gloomy
a shed as could be imagined, I felt as contented as if I
had obtained a well-furnished mansion, and looked forward
to a month's residence in it with unmixed satisfaction.

The next morning, after an early breakfast, I set off to
explore the virgin forests of Aru, anxious to set my mind
at rest as to the treasures they were likely to yield, and
the probable success of my long-meditated expedition. A
little native imp was our guide, seduced by the gift of a
German knife, value three-halfpence, and my Macassar
boy Baderoon brought his chopper to clear the path if
necessary.

We had to walk about half a mile along the beach, the
ground behind the village being mostly swampy, and then
turned into the forest along a path which leads to the
native village of Wamma, about three miles off on the
other side of the island. The path was a narrow one, and
very little used, often swampy and obstructed by fallen
trees, so that after about a mile we lost it altogether, our
guide having turned back, and we were obliged to follow
his example. In the meantime, however, I had not been
idle, and my day's captures determined the success of my

journey in an entomological point of view. I had taken
about thirty species of butterflies, more than I had ever
captured in a day since leaving the prolific banks of the
Amazon, and among them were many most rare and
beautiful insects, hitherto only known by a few specimens
from New Guinea. The large and handsome spectre-
butterfly, Hestia durvillei; the pale-winged peacock
butterfly, Drusilla catops ; and the most brilliant and
wonderful of the clear-winged moths, Cocytia durvillei,
were especially interesting, as well as several little
"blues," equalling in brilliancy and beauty anything the
butterfly world can produce.   In the other groups of
insects I was not so successful, but this was not to be
wondered at in a mere exploring ramble, when only what
is most conspicuous and novel attracts the attention.
Several pretty beetles, a superb "bug," and a few nice
land-shells were obtained, and I returned in the afternoon
well satisfied with my first trial of the promised land.

The next two days were so wet and windy that there
was no going out; but on the succeeding one the sun shone
brightly, and I had the good fortune to capture one of the
most magnificent insects the world contains, the great bird-
winged butterfly, Ornithoptera poseidon. I trembled with
excitement as I saw it coming majestically towards me,
and could hardly believe I had really succeeded in my
stroke till I had taken it out of the net and was gazing,

lost in admiration, at the velvet black and brilliant green
of its wings, seven inches across, its golden body, and
crimson breast. It is true I had seen similar insects in
cabinets at home, but it is quite another thing to capture
such oneself—to feel it struggling between one's fingers,
and to gaze upon its fresh and living beauty, a bright gem
shining out amid the silent gloom of a dark and tangled
forest. The village of Dobbo held that evening at least
one contented man.

Jan. 26th.—Having now been here a fortnight, I
began to understand a little of the place and its peculiarities. Praus continually arrived, and the merchant
population increased almost daily. Every two or three
days a fresh house was opened, and the necessary repairs
made. In every direction men were bringing in poles,
bamboos, rattans, and the leaves of the nipa palm to
construct or repair the walls, thatch, doors, and shutters of
their houses, which they do with great celerity. Some of
the arrivals were Macassar men or Bugis, but more from
the small island of Goram, at the east end of Ceram,
whose inhabitants are the petty traders of the far East.
Then the natives of Aru come in from the other side of
the islands (called here " blakang tana," or " back of the
country ") with the produce they have collected during
the preceding six months, and which they now sell to the
traders, to some of whom they are most likely in debt.

Almost all, or I may safely say all, the new arrivals pay
me a visit, to see with their own eyes the unheard-of phenomenon of a person come to stay at Dobbo who does not
trade ! They have their own ideas of the uses that may
possibly be made of stuffed birds, beetles, and shells which
are not the right shells—that is, " mother-of-pearl." They
every day bring me dead and broken shells, such as I can
pick up by hundreds on the beach, and seem quite puzzled
and distressed when I decline them. If, however, there
are any snail shells among a lot, I take them, and ask for
more—a principle of selection so utterly unintelligible to
them, that they give it up in despair, or solve the problem
by imputing hidden medical virtue to those which they
see me preserve so carefully.

These traders are all of the Malay race, or a mixture of
which Malay is the chief ingredient, with the exception of
a few Chinese. The natives of Aru, on the other hand,
are Papuans, with black or sooty brown skins, woolly
or frizzly hair, thick-ridged prominent noses, and rather
slender limbs. Most of them wear nothing but a waist-
cloth, and a few of them may be seen all day long wan-
dering about the half-deserted streets of Dobbo offering
their little bit of merchandise for sale.

Living in a trader's house everything is brought to me as
well as to the rest,—bundles of smoked tripang, or " bêche
de mer," looking like sausages which have been rolled in

mud and then thrown up the chimney ; dried sharks' fins, mother-of-pearl shells, as well as Birds of Paradise, which, however, are so dirty and so badly preserved that I have as yet found no specimens worth purchasing. When I hardly look at the articles, and make no offer for them, they seem incredulous, and, as if fearing they have misunderstood me, again offer them, and declare what they want in return —knives, or tobacco, or sago, or handkerchiefs. I then have to endeavour to explain, through any interpreter who may be at hand, that neither tripang nor pearl oyster shells have any charms for me, and that I even decline to specu- late in tortoiseshell, but that anything eatable I will buy— fish, or turtle, or vegetables of any sort. Almost the only food, however, that we can obtain with any regularity, are fish and cockles of very good quality, and to supply our daily wants it is absolutely necessary to be always pro- vided with four articles—tobacco, knives, sago-cakes, and Dutch copper doits—because when the particular thing asked for is not forthcoming, the fish pass on to the next house, and we may go that day without a dinner. It is curious to see the baskets and buckets used here. The cockles are brought in large volute shells, probably the Cymbium ducale, while gigantic helmet-shells, a species of Cassis, suspended by a rattan handle, form the vessels in which fresh water is daily carried past my door. It is painful to a naturalist to see these splendid shells with their inner whorls ruthlessly broken away to fit them for their ignoble use.

My collections, however, got on but slowly, owing to the unexpectedly bad weather, violent winds with heavy showers having been so continuous as only to give me four good collecting days out of the first sixteen I spent here. Yet enough had been collected to show me that with time and fine weather I might expect to do something good. From the natives I obtained some very fine insects and a few pretty land-shells; and of the small number of birds yet shot more than half were known New Guinea species, and therefore certainly rare in European collections, while the remainder were probably new. In one respect my hopes seemed doomed to be disappointed. I had antici- pated the pleasure of myself preparing fine specimens of the Birds of Paradise, but I now learnt that they are all at this season out of plumage, and that it is in September and October that they have the long plumes of yellow silky feathers in full perfection. As all the praus return in July, I should not be able to spend that season in Aru without remaining another whole year, which was out of the question. I was informed, however, that the small red species, the "King Bird of Paradise," retains its plumage at all seasons, and this I might therefore hope to get.

As I became familiar with the forest scenery of the island, I perceived it to possess some characteristic features that distinguished it from that of Borneo and Malacca, while, what is very singular and interesting, it recalled to my mind the half-forgotten impressions of the forests of Equatorial America. For example, the palms were much more abundant than I had generally found them in the East, more generally mingled with the other vegetation, more varied in form and aspect, and presenting some of those lofty and majestic smooth-stemmed, pinnate-leaved species which recall the Uauassu (Attalea speciosa) of the Amazon, but which I had hitherto rarely met with in the Malayan islands.

In animal life the immense number and variety of spiders and of lizards were circumstances that recalled the prolific regions of South America, more especially the abundance and varied colours of the little jumping spiders which abound on flowers and foliage, and are often perfect gems of beauty. The web-spinning species were also more numerous than I had ever seen them, and were a great annoyance, stretching their nets across the footpaths just about the height of my face; and the threads composing these are so strong and glutinous as to require much trouble to free oneself from them. Then their inhabitants, great yellow-spotted monsters with bodies two inches long, and legs in proportion, are not pleasant things to run one's nose against while pursuing some gorgeous butterfly, or gazing aloft in search of some strange-voiced bird. I soon found it necessary not only to brush away the web, but also to destroy the spinner; for at first, having cleared the path one day, I found the next morning that the industrious insects had spread their nets again in the very same places.

The lizards were equally striking by their numbers, variety, and the situations in which they were found. The beautiful blue-tailed species so abundant in Ké, was not seen here. The Aru lizards are more varied but more sombre in their colours—shades of green, grey brown, and even black, being very frequently seen. Every shrub and herbaceous plant was alive with them, every rotten trunk or dead branch served as a station for some of these active little insect-hunters, who, I fear, to satisfy their gross appetites, destroy many gems of the insect world, which would feast the eyes and delight the heart of our more discriminating entomologists. Another curious feature of the jungle here was the multitude of sea-shells everywhere met with on the ground and high up on the branches and foliage, all inhabited by hermit-crabs, who forsake the beach to wander in the forest. I have actually seen a spider carrying away a good-sized shell and devouring its (probably juvenile) tenant. On the beach, which I had to walk along every morning to reach the forest, these creatures swarmed by thousands. Every dead shell, from the largest to the most minute, was appropriated by them.

They formed small social parties of ten or twenty around
bits of stick or seaweed, but dispersed hurriedly at the
sound of approaching footsteps.   After a windy night, that
nasty-looking Chinese delicacy the sea-slug was sometimes
thrown up on the beach, which was at such times thickly
strewn with some of the most beautiful shells that adorn
our cabinets, along with fragments and masses of coral
and strange sponges, of which I picked up more than
twenty different sorts.   In many cases sponge and coral
are so much alike that it is only on touching them that
they can be distinguished.   Quantities of seaweed, too,
are thrown up; but strange as it may seem, these are far
less beautiful and less varied than may be found on any
favourable part of our own coasts.

The natives here, even those who seem to be of pure
Papuan race, were much more reserved and taciturn than
those of Ke.   This is probably because I only saw them
as yet among strangers and in small parties.   One must
see the savage at home to know what he really is.   Even
here, however the Papuan character sometimes breaks out.
Little boys sing cheerfully as they walk along, or talk
aloud to themselves (quite a negro characteristic); and, try
all they can, the men cannot conceal their emotions in the
true Malay fashion.   A number of them were one day in
my house, and having a fancy to try what sort of eating
tripang would be, I bought a couple, paying for them with

such an extravagant quantity of tobacco that the seller
saw I was a green customer. He could not, however,
conceal his delight, but as he smelt the fragrant weed, and
exhibited the large handful to his companions, he grinned
and twisted and gave silent chuckles in a most expressive
pantomime. I had often before made the same mistake in
paying a Malay for some trifle. In no case, however, was
his pleasure visible on his countenance—a dull and stupid
hesitation only showing his surprise, which would be
exhibited exactly in the same way whether he was over
or under paid. These little moral traits are of the greatest
interest when taken in connexion with physical features.
They do not admit of the same ready explanation by
external causes which is so frequently applied to the
latter. Writers on the races of mankind have too often
to trust to the information of travellers who pass rapidly
from country to country, and thus have few opportunities
of becoming acquainted with peculiarities of national character, or even of ascertaining what is really the average
physical conformation of the people. Such are exceedingly apt to be deceived in places where two races have
long intermingled, by looking on intermediate forms and
mixed habits as evidences of a natural transition from one
race to the other, instead of an artificial mixture of two
distinct peoples; and they will be the more readily led
into this error if, as in the present case, writers on the

subject should have been in the habit of classing these
races as mere varieties of one stock, as closely related in
physical conformation as from their geographical proximity
one might suppose they ought to be   So far as I have yet
seen, the Malay and Papuan appear to be as widely separated as any two human races that exist, being distinguished by physical, mental, and moral characteristics, all
of the most marked and striking kind.

Feb 5th.—I took advantage of a very fine calm day to
pay a visit to the island of Wokan, which is about a mile
from us, and forms part of the "tanna busar," or mainland of Aru.   This is a large island, extending from
north to south about a hundred miles, but so low in many
parts as to be intersected by several creeks, which run
completely through it, offering a passage for good-sized
vessels.   On the west side, where we are, there are only a
few outlying islands, of which ours (Wamma) is the
principal; but on the east coast are a great number of
islands, extending some miles beyond the mainland, and
forming the "blakang tana," or "back country," of the
traders, being the principal seat of the pearl, tripang, and
tortoiseshell fisheries.   To the mainland many of the
birds and animals of the country are altogether confined;
the Birds of Paradise, the black cockatoo, the great brush-
turkey, and the cassowary, are none of them found on
Wamma or any of the detached islands.   I did not,

however, expect in this excursion to see any decided difference in the forest or its productions, and was therefore
agreeably surprised. The beach was overhung with the
drooping branches of large trees, loaded with Orchideæ,
ferns, and other epiphytal plants. In the forest there was
more variety, some parts being dry, and with trees of a
lower growth, while in others there were some of the most
beautiful palms I have ever seen, with a perfectly straight,
smooth, slender stem, a hundred feet high, and a crown of
handsome drooping leaves. But the greatest novelty and
most striking feature to my eyes were the tree-ferns, which,
after seven years spent in the tropics, I now saw in perfection for the first time. All I had hitherto met with
were slender species, not more than twelve feet high, and
they gave not the least idea of the supreme beauty of trees
bearing their elegant heads of fronds more than thirty feet
in the air, like those which were plentifully scattered about
this forest.   There is nothing in tropical vegetation so
perfectly beautiful.

My boys shot five sorts of birds, none of which we had
obtained during a month's shooting in Wamma. Two
were very pretty flycatchers, already known from New
Guinea ; one of them (Monarcha chrysomela), of brilliant
black and bright orange colours, is by some authors considered to be the most beautiful of all flycatchers; the
other is pure white and velvety black, with a broad fleshy

ring round the eye of an azure blué colour; it is named
the "spectacled flycatcher" (Monarcha telescopthalma),
and was first found in New Guinea, along with the other,
by the French naturalists during the voyage of the dis-
covery-ship *Coquille.*

Feb. 18*th.*—Before leaving Macassar, I had written to
the Governor of Amboyna requesting him to assist me
with the native chiefs of Aru. I now received by a
vessel which had arrived from Amboyna a very polite
answer, informing me that orders had been sent to give
me every assistance that I might require; and I was just
congratulating myself on being at length able to get a boat
and men to go to the mainland and explore the interior,
when a sudden check came in the form of a piratical
incursion. A small prau arrived which had been
attacked by pirates and had a man wounded. They
were said to have five boats, but more were expected to be
behind, and the traders were all in consternation, fearing
that their small vessels sent trading to the "blakang tana"
would be plundered. The Aru natives were of course
dreadfully alarmed, as these marauders attack their
villages, burn and murder, and carry away women and
children for slaves. Not a man will stir from his village
for some time, and I must remain still a prisoner in
Dobbo. The Governor of Amboyna, out of pure kindness, has told the chiefs that they are to be respon-

sible for my safety, so that they have an excellent excuse for refusing to stir.

Several praus went out in search of the pirates, sentinels were appointed, and watch-fires lighted on the beach to guard against the possibility of a night attack, though it was hardly thought they would be bold enough to attempt to plunder Dobbo. The next day the praus returned, and we had positive information that these scourges of the Eastern seas were really among us. One of Herr Warzbergen's small praus also arrived in a sad plight. It had been attacked six days before, just as it was returning from the "blakang tana." The crew escaped in their small boat and hid in the jungle, while the pirates came up and plundered the vessel. They took away everything but the cargo of mother-of-pearl shell, which was too bulky for them. All the clothes and boxes of the men, and the sails and cordage of the prau, were cleared off. They had four large war boats, and fired a volley of musketry as they came up, and sent off their small boats to the attack. After they had left, our men observed from their concealment that three had stayed behind with a small boat ; and being driven to desperation by the sight of the plundering, one brave fellow swam off armed only with his parang, or chopping-knife, and coming on them unawares made a desperate attack, killing one and wounding the other two, receiving himself numbers of slight wounds,

and then swimming off again when almost exhausted.
Two other praus were also plundered, and the crew of one
of them murdered to a man. They are said to be Sooloo
pirates, but have Bugis among them. On their way here
they have devastated one of the small islands east of
Ceram.   It is now eleven years since they have visited
Aru, and by thus making their attacks at long and uncertain intervals the alarm dies away, and they find a
population for the most part unarmed and unsuspicious of
danger.   None of the small trading vessels now carry
arms, though they did so for a year or two after the last
attack, which was just the time when there was the least
occasion for it.   A week later one of the smaller pirate
boats was captured in the " blakang tana."   Seven men
were killed and three taken prisoners. The larger vessels
have been often seen but cannot be caught, as they have
very strong crews, and can always escape by rowing out
to sea in the eye of the wind, returning at night.   They
will thus remain among the innumerable islands and
channels, till the change of the monsoon enables them to
sail westward.

March 9th.—For four or five days we have had a continual gale of wind, with occasional gusts of great fury,
which seem as if they would send Dobbo into the sea.
Rain accompanies it almost every alternate hour, so that
it is not a pleasant time.   During such weather I can do

little, but am busy getting ready a boat I have purchased, for an excursion into the interior. There is immense difficulty about men, but I believe the "Orang-kaya," or head man of Wamma, will accompany me to see that I don't run into danger.

Having become quite an old inhabitant of Dobbo, I will endeavour to sketch the sights and sounds that pervade it, and the manners and customs of its inhabitants. The place is now pretty full, and the streets present a far more cheerful aspect than when we first arrived. Every house is a store, where the natives barter their produce for what they are most in need of. Knives, choppers, swords, guns, tobacco, gambier, plates, basins, handkerchiefs, sarongs, calicoes, and arrack, are the principal articles wanted by the natives ; but some of the stores contain also tea, coffee, sugar, wine, biscuits, &c., for the supply of the traders; and others are full of fancy goods, china ornaments, looking-glasses, razors, umbrellas, pipes, and purses, which take the fancy of the wealthier natives. Every fine day mats are spread before the doors and the tripang is put out to dry, as well as sugar, salt, biscuit, tea, cloths, and other things that get injured by an excessively moist atmosphere. In the morning and evening, spruce Chinamen stroll about or chat at each other's doors, in blue trousers, white jacket, and a queue into which red silk is plaited till it reaches almost to their heels. An old Bugis hadji regularly takes

an evening stroll in all the dignity of flowing green silk
robe and gay turban, followed by two small boys carrying
his sirih and betel boxes.

In every vacant space new houses are being built, and
all sorts of odd little cooking-sheds are erected against the
old ones, while in some out-of-the-way corners, massive log
pigsties are tenanted by growing porkers; for how could
the Chinamen exist six months without one feast of pig?
Here and there are stalls where bananas are sold, and
every morning two little boys go about with trays of sweet
rice and grated cocoa-nut, fried fish, or fried plantains; and
whichever it may be, they have but one cry, and that is—
" Chocolat—t—t !"   This must be a Spanish or Portuguese
cry, handed down for centuries, while its meaning has
been lost.   The Bugis sailors, while hoisting the mainsail, cry out, " Vēla à vēla,—véla, véla, vēla !" repeated
in an everlasting chorus.   As " vela " is Portuguese
for a sail, I supposed I had discovered the origin of
this, but I found afterwards they used the same cry
when heaving anchor, and often changed it to "hela,"
which is so much an universal expression of exertion
and hard breathing that it is most probably a mere in-
terjectional cry.

I daresay there are now near five hundred people in
Dobbo of various races, all met in this remote corner of
the East, as they express it, " to look for their fortune ; " to

get money any way they can. They are most of them
people who have the very worst reputation for honesty as
well as every other form of morality,—Chinese, Bugis,
Ceramese, and half-caste Javanese, with a sprinkling of
half-wild Papuans from Timor, Babber, and other islands,—
yet all goes on as yet very quietly. This motley, ignorant,
bloodthirsty, thievish population live here without the
shadow of a government, with no police, no courts, and no
lawyers ; yet they do not cut each other's throats, do not
plunder each other day and night, do not fall into the
anarchy such a state of things might be supposed to lead
to. It is very extraordinary ! It puts strange thoughts
into one's head about the mountain-load of government
under which people exist in Europe, and suggests the idea
that we may be overgoverned. Think of the hundred
Acts of Parliament annually enacted to prevent us, the
people of England, from cutting each other's throats,
or from doing to our neighbour as we would *not* be
done by. Think of the thousands of lawyers and bar-
risters whose whole lives are spent in telling us what
the hundred Acts of Parliament mean, and one would be
led to infer that if Dobbo has too little law England has
too much.

Here we may behold in its simplest form the genius of
Commerce at the work of Civilization.   Trade is the magic
that keeps all at peace, and unites these discordant elements

into a well-behaved community. All are traders, and all
know that peace and order are essential to successful trade,
and thus a public opinion is created which puts down all
lawlessness. Often in former years, when strolling along
the Campong Glam in Singapore, I have thought how wild
and ferocious the Bugis sailors looked, and how little I
should like to trust myself among them. But now I find
them to be very decent, well-behaved fellows; I walk
daily unarmed in the jungle, where I meet them continually; I sleep in a palm-leaf hut, which any one may
enter, with as little fear and as little danger of thieves or
murder as if I were under the protection of the Metro-
politan police. It is true the Dutch influence is felt here.
The islands are nominally under the government of the
Moluccas, which the native chiefs acknowledge; and in
most years a commissioner arrives from Amboyna, who
makes the tour of the islands, hears complaints, settles
disputes, and carries away prisoner any heinous offender.
This year he is not expected to come, as no orders have yet
been received to prepare for him; so the people of Dobbo
will probably be left to their own devices. One day a
man was caught in the act of stealing a piece of iron from
Herr Warzbergen's house, which he had entered by making
a hole through the thatch wall. In the evening the chief
traders of the place, Bugis and Chinese, assembled, the
offender was tried and found guilty, and sentenced to

receive twenty lashes on the spot.   They were given with
a small rattan in the middle of the street, not very severely,
as the executioner appeared to sympathise a little with the
culprit.   The disgrace seemed to be thought as much of as
the pain; for though any amount of clever cheating is
thought rather meritorious than otherwise, open robbery
and housebreaking meet with universal reprobation.

CHAPTER XXXI.

THE ARU ISLANDS.—JOURNEY AND RESIDENCE IN THE INTERIOR.

(MARCH TO MAY 1857.)

MY boat was at length ready, and having obtained two men besides my own servants, after an enormous amount of talk and trouble, we left Dobbo on the morning of March 13th, for the mainland of Aru. By noon we reached the mouth of a small river or creek, which we ascended, winding among mangrove swamps, with here and there a glimpse of dry land. In two hours we reached a house, or rather small shed, of the most miserable description, which our steersman, the "Orang-kaya" of Wamma, said was the place we were to stay at, and where he had assured me we could get every kind of bird and beast to be found in Aru. The shed was occupied by about a dozen men, women, and children; two cooking fires were burning in it, and there seemed little prospect of my obtaining any accommodation. I however deferred

inquiry till I had seen the neighbouring forest, and immediately started off with two men, net, and guns, along a
path at the back of the house.   In an hour's walk I saw

MAP
OF THE
ARU ISLANDS.

Mr. Wallace's Routes.

enough to make me determine to give the place a trial, and
on my return, finding the "Orang-kaya" was in a strong
fever-fit and unable to do anything, I entered into negotiations with the owner of the house for the use of a slip
at one end of it about five feet wide, for a week, and agreed
to pay as rent one "parang," or chopping-knife. I then
immediately got my boxes and bedding out of the boat,
hung up a shelf for my bird-skins and insects, and got all
ready for work next morning. My own boys slept in the
boat to guard the remainder of my property; a cooking
place sheltered by a few mats was arranged under a tree
close by, and I felt that degree of satisfaction and enjoyment which I always experience when, after much trouble
and delay, I am on the point of beginning work in a new
locality.

One of my first objects was to inquire for the people
who are accustomed to shoot the Paradise birds. They
lived at some distance in the jungle, and a man was sent
to call them. When they arrived, we had a talk by means
of the "Orang-kaya" as interpreter, and they said they
thought they could get some. They explained that they
shoot the birds with a bow and arrow, the arrow having a
conical wooden cap fitted to the end as large as a teacup,
so as to kill the bird by the violence of the blow without
making any wound or shedding any blood. The trees
frequented by the birds are very lofty; it is therefore

necessary to erect a small leafy covering or hut among the branches, to which the hunter mounts before daylight in the morning and remains the whole day, and whenever a bird alights they are almost sure of securing it. (See Frontispiece.) They returned to their homes the same evening, and I never saw anything more of them, owing, as I afterwards found, to its being too early to obtain birds in good plumage.

The first two or three days of our stay here were very wet, and I obtained but few insects or birds, but at length, when I was beginning to despair, my boy Baderoon returned one day with a specimen which repaid me for months of delay and expectation. It was a small bird, a little less than a thrush. The greater part of its plumage was of an intense cinnabar red, with a gloss as of spun glass. On the head the feathers became short and velvety, and shaded into rich orange. Beneath, from the breast downwards, was pure white, with the softness and gloss of silk, and across the breast a band of deep metallic green separated this colour from the red of the throat. Above each eye was a round spot of the same metallic green; the bill was yellow, and the feet and legs were of a fine cobalt blue, strikingly contrasting with all the other parts of the body. Merely in arrangement of colours and texture of plumage this little bird was a gem of the first water, yet these comprised only half its strange beauty. Springing

from each side of the breast, and ordinarily lying concealed
under the wings, were little tufts of greyish feathers about
two inches long, and each terminated by a broad band of
intense emerald green. These plumes can be raised at the
will of the bird, and spread out into a pair of elegant fans
when the wings are elevated. But this is not the only
ornament. The two middle feathers of the tail are in the
form of slender wires about five inches long, and which
diverge in a beautiful double curve. About half an inch of
the end of this wire is webbed on the outer side only, and
coloured of a fine metallic green, and being curled spirally
inwards form a pair of elegant glittering buttons, hanging
five inches below the body, and the same distance apart.
These two ornaments, the breast fans and the spiral
tipped tail wires, are altogether unique, not occurring on
any other species of the eight thousand different birds
that are known to exist upon the earth; and, combined
with the most exquisite beauty of plumage, render this
one of the most perfectly lovely of the many lovely productions of nature. My transports of admiration and
delight quite amused my Aru hosts, who saw nothing
more in the " Burong raja" than we do in the robin or
the goldfinch.[1]

Thus one of my objects in coming to the far East was

[1] See the upper figure on Plate VIII. at commencement of Chapter
XXXVIII.

accomplished. I had obtained a specimen of the King
Bird of Paradise (Paradisea regia), which had been described by Linnæus from skins preserved in a mutilated
state by the natives. I knew how few Europeans had
ever beheld the perfect little organism I now gazed upon,
and how very imperfectly it was still known in Europe.
The emotions excited in the minds of a naturalist, who has
long desired to see the actual thing which he has hitherto
known only by description, drawing, or badly-preserved
external covering—especially when that thing is of surpassing rarity and beauty, require the poetic faculty fully
to express them. The remote island in which I found
myself situated, in an almost unvisited sea, far from the
tracks of merchant fleets and navies; the wild luxuriant
tropical forest, which stretched far away on every side ;
the rude uncultured savages who gathered round me,—all
had their influence in determining the emotions with which
I gazed upon this "thing of beauty." I thought of the
long ages of the past, during which the successive generations of this little creature had run their course—year
by year being born, and living and dying amid these
dark and gloomy woods, with no intelligent eye to gaze
upon their loveliness; to all appearance such a wanton
waste of beauty. Such ideas excite a feeling of melancholy. It seems sad, that on the one hand such exquisite
creatures should live out their lives and exhibit their

charms only in these wild inhospitable regions, doomed
for ages yet to come to hopeless barbarism; while on the
other hand, should civilized man ever reach these distant
lands, and bring moral, intellectual, and physical light
into the recesses of these virgin forests, we may be sure
that he will so disturb the nicely-balanced relations of
organic and inorganic nature as to cause the disappearance,
and finally the extinction, of these very beings whose
wonderful structure and beauty he alone is fitted to appreciate and enjoy. This consideration must surely tell us
that all living things were *not* made for man. Many of
them have no relation to him. The cycle of their existence has gone on independently of his, and is disturbed or
broken by every advance in man's intellectual development; and their happiness and enjoyments, their loves
and hates, their struggles for existence, their vigorous life
and early death, would seem to be immediately related to
their own well-being and perpetuation alone, limited only
by the equal well-being and perpetuation of the number-
less other organisms with which each is more or less inti-
mately connected.

After the first king-bird was obtained, I went with my
men into the forest, and we were not only rewarded with
another in equally perfect plumage, but I was enabled to
see a little of the habits of both it and the larger species.
It frequents the lower trees of the less dense forests, and is

very active, flying strongly with a whirring sound, and continually hopping or flying from branch to branch. It eats hard stone-bearing fruits as large as a gooseberry, and often flutters its wings after the manner of the South American manakins, at which time it elevates and expands the beautiful fans with which its breast is adorned. The natives of Aru call it "Goby-goby."

One day I got under a tree where a number of the Great Paradise birds were assembled, but they were high up in the thickest of the foliage, and flying and jumping about so continually that I could get no good view of them. At length I shot one, but it was a young specimen, and was entirely of a rich chocolate-brown colour, without either the metallic green throat or yellow plumes of the full-grown bird. All that I had yet seen resembled this, and the natives told me that it would be about two months before any would be found in full plumage. I still hoped, therefore, to get some. Their voice is most extraordinary. At early morn, before the sun has risen, we hear a loud cry of "Wawk—wawk—wawk, wŏk—wŏk—wŏk," which resounds through the forest, changing its direction continually. This is the Great Bird of Paradise going to seek his breakfast. Others soon follow his example ; lories and parroquets cry shrilly, cockatoos scream, king-hunters croak and bark, and the various smaller birds chirp and whistle their morning song. As I lie listening to these

interesting sounds, I realize my position as the first
European who has ever lived for months together in the
Aru islands, a place which I had hoped rather than
expected ever to visit. I think how many besides myself
have longed to reach these almost fairy realms, and to see
with their own eyes the many wonderful and beautiful
things which I am daily encountering. But now Ali
and Baderoon are up and getting ready their guns and
ammunition, and little Baso has his fire lighted and is
boiling my coffee, and I remember that I had a black
cockatoo brought in late last night, which I must skin
immediately, and so I jump up and begin my day's work
very happily.

This cockatoo is the first I have seen, and is a great
prize. It has a rather small and weak body, long weak
legs, large wings, and an enormously developed head,
ornamented with a magnificent crest, and armed with a
sharp-pointed hooked bill of immense size and strength.
The plumage is entirely black, but has all over it the
curious powdery white secretion characteristic of cockatoos.
The cheeks are bare, and of an intense blood-red colour.
Instead of the harsh scream of the white cockatoos, its
voice is a somewhat plaintive whistle. The tongue is a
curious organ, being a slender fleshy cylinder of a deep
red colour, terminated by a horny black plate, furrowed
across and somewhat prehensile. The whole tongue has

a considerable extensile power. I will here relate something of the habits of this bird, with which I have since
become acquainted. It frequents the lower parts of the
forest, and is seen singly, or at most two or three together.
It flies slowly and noiselessly, and may be killed by a
comparatively slight wound. It eats various fruits and
seeds, but seems more particularly attached to the kernel
of the kanary-nut, which grows on a lofty forest tree
(Canarium commune), abundant in the islands where this
bird is found ; and the manner in which it gets at these
seeds shows a correlation of structure and habits, which
would point out the "kanary" as its special food. The
shell of this nut is so excessively hard that only a heavy
hammer will crack it; it is somewhat triangular, and the
outside is quite smooth. The manner in which the bird
opens these nuts is very curious. Taking one endways in
its bill and keeping it firm by a pressure of the tongue, it
cuts a transverse notch by a lateral sawing motion of the
sharp-edged lower mandible. This done, it takes hold of
the nut with its foot, and biting off a piece of leaf retains
it in the deep notch of the upper mandible, and again
seizing the nut, which is prevented from slipping by the
elastic tissue of the leaf, fixes the edge of the lower
mandible in the notch, and by a powerful nip breaks off
a piece of the shell. Again taking the nut in its claws,
it inserts the very long and sharp point of the bill and

HEAD OF BLACK COCKATOO.

picks out the kernel, which is seized hold of, morsel by
morsel, by the extensible tongue. Thus every detail of
form and structure in the extraordinary bill of this bird

seems to have its use, and we may easily conceive that the black cockatoos have maintained themselves in competition with their more active and more numerous white allies, by their power of existing on a kind of food which no other bird is able to extract from its stony shell. The species is the Microglossum aterrimum of naturalists.

During the two weeks which I spent in this little settlement, I had good opportunities of observing the natives at their own home, and living in their usual manner. There is a great monotony and uniformity in every-day savage life, and it seemed to me a more miserable existence than when it had the charm of novelty. To begin with the most important fact in the existence of uncivilized peoples—their food—the Aru men have no regular supply, no staff of life, such as bread, rice, mandiocca, maize, or sago, which are the daily food of a large proportion of mankind. They have, however, many sorts of vegetables, plantains, yams, sweet potatoes, and raw sago; and they chew up vast quantities of sugar-cane, as well as betel-nuts, gambir, and tobacco. Those who live on the coast have plenty of fish; but when inland, as we are here, they only go to the sea occasionally, and then bring home cockles and other shell-fish by the boatload. Now and then they get wild pig or kangaroo, but too rarely to form anything like a regular part of their diet, which is essentially vegetable ; and what is of more importance, as affecting their health, green, watery vegetables, imperfectly cooked, and even these in varying and often insufficient quantities. To this diet may be attributed the prevalence of skin diseases, and ulcers on the legs and joints. The scurfy skin disease so common among savages has a close connexion with the poorness and irregularity of their living. The Malays, who are never without their daily rice, are generally free from it; the hill-Dyaks of Borneo, who grow rice and live well, are clean skinned, while the less industrious and less cleanly tribes, who live for a portion of the year on fruits and vegetables only, are very subject to this malady. It seems clear that in this, as in other respects, man is not able to make a beast of himself with impunity, feeding like the cattle on the herbs and fruits of the earth, and taking no thought of the morrow. To maintain his health and beauty he must labour to prepare some farinaceous product capable of being stored and accumulated, so as to give him a regular supply of wholesome food. When this is obtained, he may add vegetables, fruits, and meat with advantage.

The chief luxury of the Aru people, besides betel and tobacco, is arrack (Java rum), which the traders bring in great quantities and sell very cheap. A day's fishing or rattan cutting will purchase at least a half-gallon bottle; and when the tripang or birds' nests collected during a season are sold, they get whole boxes,

each containing fifteen such bottles, which the inmates
of a house will sit round day and night till they have
finished. They themselves tell me that at such bouts they
often tear to pieces the house they are in, break and
destroy everything they can lay their hands on, and make
such an infernal riot as is alarming to behold.

The houses and furniture are on a par with the food.
A rude shed, supported on rough and slender sticks rather
than posts, no walls, but the floor raised to within a foot
of the eaves, is the style of architecture they usually
adopt. Inside there are partition walls of thatch, forming
little boxes or sleeping places, to accommodate the two or
three separate families that usually live together. A few
mats, baskets, and cooking vessels, with plates and basins
purchased from the Macassar traders, constitute their
whole furniture; spears and bows are their weapons; a
sarong or mat forms the clothing of the women, a waist-
cloth of the men. For hours or even for days they sit
idle in their houses, the women bringing in the vegetables
or sago which form their food. Sometimes they hunt or
fish a little, or work at their houses or canoes, but they
seem to enjoy pure idleness, and work as little as they
can. They have little to vary the monotony of life, little
that can be called pleasure, except idleness and conversation. And they certainly do talk! Every evening there
is a little Babel around me: but as I understand not a

word of it, I go on with my book or work undisturbed.
Now and then they scream and shout, or laugh frantically
for variety ; and this goes on alternately with vociferous
talking of men, women, and children, till long after I am
in my mosquito curtain and sound asleep.

At this place I obtained some light on the complicated
mixture of races in Aru, which would utterly confound an
ethnologist.   Many of the natives, though equally dark
with the others, have little of the Papuan physiognomy,
but have more delicate features of the European type,
with more glossy, curling hair. These at first quite puzzled
me, for they have no more resemblance to Malay than to
Papuan, and the darkness of skin and hair would forbid
the idea of Dutch intermixture.   Listening to their conversation, however, I detected some words that were
familiar to me.   " Accabó" was one ; and to be sure that
it was not an accidental resemblance, I asked the speaker
in Malay what "accabó" meant, and was told it meant
"done or finished," a true Portuguese word, with its
meaning retained.   Again, I heard the word "jafui" often
repeated, and could see, without inquiry, that its meaning
was "he's gone," as in Portuguese.   "Porco," too, seems
a common name, though the people have no idea of its
European meaning. This cleared up the difficulty. I at
once understood that some early Portuguese traders had
penetrated to these islands, and mixed with the natives,

influencing their language, and leaving in their descendants for many generations the visible characteristics of their race. If to this we add the occasional mixture of Malay, Dutch, and Chinese with the indigenous Papuans, we have no reason to wonder at the curious varieties of form and feature occasionally to be met with in Aru. In this very house there was a Macassar man, with an Aru wife and a family of mixed children. In Dobbo I saw a Javanese and an Amboyna man, each with an Aru wife and family; and as this kind of mixture has been going on for at least three hundred years, and probably much longer, it has produced a decided effect on the physical characteristics of a considerable portion of the population of the islands, more especially in Dobbo and the parts nearest to it.

March 28*th.*—The "Orang-kaya" being very ill with fever had begged to go home, and had arranged with one of the men of the house to go on with me as his substitute. Now that I wanted to move, the bugbear of the pirates was brought up, and it was pronounced unsafe to go further than the next small river. This would not suit me, as I had determined to traverse the channel called Watelai to the "blakang-tana;" but my guide was firm in his dread of pirates, of which I knew there was now no danger, as several vessels had gone in search of them, as well as a Dutch gunboat which had arrived since I left Dobbo. I had, fortunately, by this time heard that the Dutch "Com-

missie" had really arrived, and therefore threatened that if
my guide did not go with me immediately, I would appeal
to the authorities, and he would certainly be obliged to give
back the cloth which the "Orang-kaya" had transferred
to him in prepayment. This had the desired effect; matters
were soon arranged, and we started the next morning.
The wind, however, was dead against us, and after rowing
hard till midday we put in to a small river where there were
a few huts, to cook our dinners. The place did not look
very promising, but as we could not reach our destination,
the Watelai river, owing to the contrary wind, I thought we
might as well wait here a day or two. I therefore paid a
chopper for the use of a small shed, and got my bed and
some boxes on shore. In the evening, after dark, we were
suddenly alarmed by the cry of "Bajak! bajak!" (Pirates!)
The men all seized their bows and spears, and rushed down
to the beach; we got hold of our guns and prepared for
action, but in a few minutes all came back laughing and
chattering, for it had proved to be only a small boat and
some of their own comrades returned from fishing. When
all was quiet again, one of the men, who could speak a
little Malay, came to me and begged me not to sleep too
hard. "Why?" said I. "Perhaps the pirates may really
come," said he very seriously, which made me laugh and
assure him I should sleep as hard as I could.

Two days were spent here, but the place was unproductive of insects or birds of interest, so we made another attempt to get on. As soon as we got a little away from the land we had a fair wind, and in six hours' sailing reached the entrance of the Watelai channel, which divides the most northerly from the middle portion of Aru. At its mouth this was about half a mile wide, but soon narrowed, and a mile or two on it assumed entirely the aspect of a river about the width of the Thames at London, winding among low but undulating and often hilly country. The scene was exactly such as might be expected in the interior of a continent. The channel continued of a uniform average width, with reaches and sinuous bends, one bank being often precipitous, or even forming vertical cliffs, while the other was flat and apparently alluvial ; and it was only the pure salt-water, and the absence of any stream but the slight flux and reflux of the tide, that would enable a person to tell that he was navigating a strait and not a river. The wind was fair, and carried us along, with occasional assistance from our oars, till about three in the afternoon, when we landed where a little brook formed two or three basins in the coral rock, and then fell in a miniature cascade into the salt water river. Here we bathed and cooked our dinner. and enjoyed ourselves lazily till sunset, when we pursued our way for two hours more, and then moored our little vessel to an overhanging tree for the night.

At five the next morning we started again, and in
an hour overtook four large praus containing the "Com-
missie," who had come from Dobbo to make their
official tour round the islands, and had passed us in the
night. I paid a visit to the Dutchmen, one of whom
spoke a little English, but we found that we could get
on much better with Malay. They told me that they
had been delayed going after the pirates to one of the
northern islands, and had seen three of their vessels but
could not catch them, because on being pursued they
rowed out in the wind's eye, which they are enabled to do
by having about fifty oars to each boat. Having had some
tea with them, I bade them adieu, and turned up a narrow
channel which our pilot said would take us to the village
of Watelai, on the west side of Aru. After going some
miles we found the channel nearly blocked up with coral,
so that our boat grated along the bottom, crunching what
may truly be called the living rock. Sometimes all hands
had to get out and wade, to lighten the vessel and lift it
over the shallowest places; but at length we overcame all
obstacles and reached a wide bay or estuary studded with
little rocks and islets, and opening to the western sea and
the numerous islands of the "blakang-tana." I now found
that the village we were going to was miles away; that we
should have to go out to sea, and round a rocky point. A
squall seemed coming on, and as I have a horror of small

boats at sea, and from all I could learn Watelai village
was not a place to stop at (no Birds of Paradise being
found there), I determined to return and go to a village
I had heard of up a tributary of the Watelai river, and
situated nearly in the centre of the mainland of Aru. The
people there were said to be good, and to be accustomed to
hunting and bird-catching, being too far inland to get any
part of their food from the sea. While I was deciding
this point the squall burst upon us, and soon raised a
rolling sea in the shallow water, which upset an oil bottle
and a lamp, broke some of my crockery, and threw us all
into confusion. Rowing hard we managed to get back
into the main river by dusk, and looked out for a place to
cook our suppers. It happened to be high water, and a
very high tide, so that every piece of sand or beach was
covered, and it was with the greatest difficulty, and after
much groping in the dark, that we discovered a little
sloping piece of rock about two feet square on which to
make a fire and cook some rice. The next day we continued our way back, and on the following day entered a
stream on the south side of the Watelai river, and ascending to where navigation ceased found the little village of
Wanumbai, consisting of two large houses surrounded by
plantations, amid the virgin forests of Aru.

As I liked the look of the place, and was desirous of
staying some time, I sent my pilot to try and make a

bargain for house accommodation. The owner and chief man of the place made many excuses. First, he was afraid I would not like his house, and then was doubtful whether his son, who was away, would like his admitting me. I had a long talk with him myself, and tried to explain what I was doing, and how many things I would buy of them, and showed him my stock of beads, and knives, and cloth, and tobacco, all of which I would spend with his family and friends if he would give me house-room. He seemed a little staggered at this, and said he would talk to his wife, and in the meantime I went for a little walk to see the neighbourhood. When I came back, I again sent my pilot, saying that I would go away if he would not give me part of his house. In about half an hour he returned with a demand for about half the cost of building a house, for the rent of a small portion of it for a few weeks. As the only difficulty now was a pecuniary one, I got out about ten yards of cloth, an axe, with a few beads and some tobacco, and sent them as my final offer for the part of the house which I had before pointed out. This was accepted after a little more talk, and I immediately proceeded to take possession.

The house was a good large one, raised as usual about seven feet on posts, the walls about three or four feet more, with a high-pitched roof. The floor was of bamboo laths, and in the sloping roof was an immense shutter,

which could be lifted and propped up to admit light and
air. At the end where this was situated the floor was
raised about a foot, and this piece, about ten feet wide
by twenty long, quite open to the rest of the house, was
the portion I was to occupy. At one end of this piece,
separated by a thatch partition, was a cooking place, with
a clay floor and shelves for crockery. At the opposite end
I had my mosquito curtain hung, and round the walls we
arranged my boxes and other stores, fitted up a table and
seat, and with a little cleaning and dusting made the place
look quite comfortable. My boat was then hauled up on
shore, and covered with palm-leaves, the sails and oars
brought indoors, a hanging-stage for drying my specimens
erected outside the house and another inside, and my
boys were set to clean their guns and get all ready for
beginning work.

The next day I occupied myself in exploring the paths
in the immediate neighbourhood. The small river up
which we had ascended ceases to be navigable at this
point, above which it is a little rocky brook, which quite
dries up in the hot season. There was now, however, a
fair stream of water in it; and a path which was partly
in and partly by the side of the water, promised well
for insects, as I here saw the magnificent blue butterfly, Papilio ulysses, as well as several other fine species,
flopping lazily along, sometimes resting high up on the

foliage which drooped over the water, at others settling down on the damp rock or on the edges of muddy pools. A little way on several paths branched off through patches of second-growth forest to cane-fields, gardens, and scattered houses, beyond which again the dark wall of verdure striped with tree-trunks, marked out the limits of the primeval forests. The voices of many birds promised good shooting, and on my return I found that my boys had already obtained two or three kinds I had not seen before; and in the evening a native brought me a rare and beautiful species of ground-thrush (Pitta novæ-guineæ) hitherto only known from New Guinea.

As I improved my acquaintance with them I became much interested in these people, who are a fair sample of the true savage inhabitants of the Aru Islands, tolerably free from foreign admixture. The house I lived in contained four or five families, and there were generally from six to a dozen visitors besides. They kept up a continual row from morning till night—talking, laughing, shouting, without intermission—not very pleasant, but interesting as a study of national character. My boy Ali said to me, "Banyak quot bitchara Orang Aru" (The Aru people are very strong talkers), never having been accustomed to such eloquence either in his own or any other country he had hitherto visited. Of an evening the men, having got over their first shyness, began to talk to me a little, asking

about my country, &c., and in return I questioned them
about any traditions they had of their own origin. I had,
however, very little success, for I could not possibly make
them understand the simple question of where the Aru
people first came from. I put it in every possible way to
them, but it was a subject quite beyond their speculations;
they had evidently never thought of anything of the kind,
and were unable to conceive a thing so remote and so
unnecessary to be thought about, as their own origin.
Finding this hopeless, I asked if they knew when the
trade with Aru first began, when the Bugis and Chinese
and Macassar men first came in their praus to buy tripang
and tortoise-shell, and birds' nests, and Paradise birds?
This they comprehended, but replied that there had always
been the same trade as long as they or their fathers recol-
lected, but that this was the first time a real white man
had come among them, and, said they, "You see how the
people come every day from all the villages round to look
at you." This was very flattering, and accounted for the
great concourse of visitors which I had at first imagined
was accidental. A few years before I had been one of the
gazers at the Zoolus and the Aztecs in London. Now the
tables were turned upon me, for I was to these people a
new and strange variety of man, and had the honour of
affording to them, in my own person, an attractive exhibition, gratis.

All the men and boys of Aru are expert archers, never stirring without their bows and arrows. They shoot all sorts of birds, as well as pigs and kangaroos occasionally, and thus have a tolerably good supply of meat to eat with their vegetables. The result of this better living is superior healthiness, well-made bodies, and generally clear skins. They brought me numbers of small birds in exchange for beads or tobacco, but mauled them terribly, notwithstanding my repeated instructions. When they got a bird alive they would often tie a string to its leg, and keep it a day or two, till its plumage was so draggled and dirtied as to be almost worthless. One of the first things I got from them was a living specimen of the curious and beautiful racquet-tailed kingfisher. Seeing how much I admired it, they afterwards brought me several more, which were all caught before daybreak, sleeping in cavities of the rocky banks of the stream. My hunters also shot a few specimens, and almost all of them had the red bill more or less clogged with mud and earth. This indicates the habits of the bird, which, though popularly a king-fisher, never catches fish, but lives on insects and minute shells, which it picks up in the forest, darting down upon them from its perch on some low branch. The genus Tanysiptera, to which this bird belongs, is remarkable for the enormously lengthened tail, which in all other kingfishers is small and short. Linnæus named the species known to him "the

goddess kingfisher" (Alcedo dea), from its extreme grace
and beauty, the plumage being brilliant blue and white,
with the bill red, like coral.　Several species of these interesting birds are now known, all confined within the
very limited area which comprises the Moluccas, New
Guinea, and the extreme North of Australia.　They
resemble each other so closely that several of them can
only be distinguished by careful comparison.　One of the
rarest, however, which inhabits New Guinea, is very distinct
from the rest, being bright red beneath instead of white.
That which I now obtained was a new one, and has been
named Tanysiptera hydrocharis, but in general form and
coloration it is exactly similar to the larger species found
in Amboyna, and figured at page 468 of my first volume.

New and interesting birds were continually brought in,
either by my own boys or by the natives, and at the end of
a week Ali arrived triumphant one afternoon with a fine
specimen of the Great Bird of Paradise.　The ornamental
plumes had not yet attained their full growth, but the
richness of their glossy orange colouring, and the exquisite
delicacy of the loosely waving feathers, were unsurpassable
At the same time a great black cockatoo was brought in, as
well as a fine fruit-pigeon and several small birds, so that
we were all kept hard at work skinning till sunset.　Just
as we had cleared away and packed up for the night, a
strange beast was brought, which had been shot by the

natives. It resembled in size, and in its white woolly covering, a small fat lamb, but had short legs, hand-like feet with large claws, and a long prehensile tail. It was a Cuscus (C. maculatus), one of the curious marsupial animals of the Papuan region, and I was very desirous to obtain the skin. The owners, however, said they wanted to eat it; and though I offered them a good price, and promised to give them all the meat, there was great hesitation. Suspecting the reason, I offered, though it was night, to set to work immediately and get out the body for them, to which they agreed. The creature was much hacked about, and the two hind feet almost cut off, but it was the largest and finest specimen of the kind I had seen; and after an hour's hard work I handed over the body to the owners, who immediately cut it up and roasted it for supper.

As this was a very good place for birds, I determined to remain a month longer, and took the opportunity of a native boat going to Dobbo, to send Ali for a fresh supply of ammunition and provisions. They started on the 10th of April, and the house was crowded with about a hundred men, boys, women, and girls, bringing their loads of sugar-cane, plantains, sirih-leaf, yams, &c.; one lad going from each house to sell the produce and make purchases. The noise was indescribable. At least fifty of the hundred were always talking at once, and that not in the low measured tones of the apathetically polite Malay, but with loud voices, shouts, and screaming laughter, in which the women and children were even more conspicuous than the men.   It was only while gazing at me that their tongues were moderately quiet, because their eyes were fully occupied. The black vegetable soil here overlying the coral rock is very rich, and the sugar-cane was finer than any I had ever seen.   The canes brought to the boat were often ten and even twelve feet long, and thick in proportion, with short joints throughout, swelling between the knots with the abundance of the rich juice.   At Dobbo they get a high price for it, 1*d.* to 3*d.* a stick, and there is an insatiable demand among the crews of the praus and the Baba fishermen.   Here they eat it continually.   They half live on it, and sometimes feed their pigs with it. Near every house are great heaps of the refuse cane ; and large wicker-baskets to contain this refuse as it is produced form a regular part of the furniture of a house.   Whatever time of the day you enter, you are sure to find three or four people with a yard of cane in one hand, a knife in the other, and a basket between their legs, hacking, paring, chewing, and basket-filling, with a persevering assiduity which reminds one of a hungry cow grazing, or of a caterpillar eating up a leaf.

After five days' absence the boats returned from Dobbo, bringing Ali and all the things I had sent for quite safe. A large party had assembled to be ready to carry home the goods brought, among which were a good many cocoa-nuts, which are a great luxury here. It seems strange that they should never plant them; but the reason simply is, that they cannot bring their hearts to bury a good nut for the prospective advantage of a crop twelve years hence. There is also the chance of the fruits being dug up and eaten unless watched night and day. Among the things I had sent for was a box of arrack, and I was now of course besieged with requests for a little drop. I gave them a flask (about two bottles), which was very soon finished, and I was assured that there were many present who had not had a taste. As I feared my box would very soon be emptied if I supplied all their demands, I told them I had given them one, but the second they must pay for, and that afterwards I must have a Paradise bird for each flask. They immediately sent round to all the neighbouring houses, and mustered up a rupee in Dutch copper money, got their second flask, and drunk it as quickly as the first, and were then very talkative, but less noisy and importunate than I had expected. Two or three of them got round me and begged me for the twentieth time to tell them the name of my country. Then, as they could not pronounce it satisfactorily, they insisted that I was deceiving them, and that it was a name of my own invention. One funny old man, who bore a ludicrous resemblance to a friend of mine at home, was almost indignant. "Ung-

lung!" said he, "who ever heard of such a name?—ang-
lang—anger-lang—that can't be the name of your country;
you are playing with us." Then he tried to give a con-
vincing illustration. "My country is Wanumbai—any-
body can say Wanumbai. I'm an 'orang-Wanumbai;'
but, N-glung! who ever heard of such a name? Do tell
us the real name of your country, and then when you are
gone we shall know how to talk about you." To this
luminous argument and remonstrance I could oppose
nothing but assertion, and the whole party remained
firmly convinced that I was for some reason or other
deceiving them. They then attacked me on another point
—what all the animals and birds and insects and shells
were preserved so carefully for. They had often asked me
this before, and I had tried to explain to them that they
would be stuffed, and made to look as if alive, and
people in my country would go to look at them. But this
was not satisfying; in my country there must be many
better things to look at, and they could not believe I
would take so much trouble with their birds and beasts
just for people to look at. They did not want to look at
them; and we, who made calico and glass and knives, and
all sorts of wonderful things, could not want things from
Aru to look at. They had evidently been thinking about
it, and had at length got what seemed a very satisfactory
theory; for the same old man said to me, in a low mys-

terious voice, "What becomes of them when you go on to
the sea?"   "Why, they are all packed up in boxes," said
I. "What did you think became of them?"   "They all
come to life again, don't they?" said he; and though I
tried to joke it off, and said if they did we should have
plenty to eat at sea, he stuck to his opinion, and kept
repeating, with an air of deep conviction, "Yes, they all
come to life again, that's what they do—they all come to
life again."

After a little while, and a good deal of talking among
themselves, he began again—"I know all about it—oh,
yes! Before you came we had rain every day—very wet
indeed; now, ever since you have been here, it is fine hot
weather. Oh, yes! I know all about it; you can't deceive
me."   And so I was set down as a conjurer, and was
unable to repel the charge. But the conjurer was completely puzzled by the next question: "What," said
the old man, "is the great ship, where the Bugis and
Chinamen go to sell their things? It is always in the great
sea—its name is Jong; tell us all about it."   In vain I
inquired what they knew about it; they knew nothing
but that it was called "Jong," and was always in the sea,
and was a very great ship, and concluded with, "Perhaps
that is your country?"   Finding that I could not or
would not tell them anything about "Jong," there came
more regrets that I would not tell them the real name of

my country ; and then a long string of compliments, to the
effect that I was a much better sort of a person than the
Bugis and Chinese, who sometimes came to trade with
them, for I gave them things for nothing, and did not try
to cheat them. How long would I stop ? was the next
earnest inquiry. Would I stay two or three months ?
They would get me plenty of birds and animals, and I
might soon finish all the goods I had brought, and then,
said the old spokesman, " Don't go away, but send for
more things from Dobbo, and stay here a year or two."
And then again the old story, " Do tell us the name of
your country. We know the Bugis men, and the Macassar
men, and the Java men, and the China men ; only you, we
don't know from what country you come. Ung-lung ! it
can't be ; I know that is not the name of your country."
Seeing no end to this long talk, I said I was tired, and
wanted to go to sleep ; so after begging—one a little bit of
dry fish for his supper, and another a little salt to eat with
his sago—they went off very quietly, and I went outside
and took a stroll round the house by moonlight, thinking of
the simple people and the strange productions of Aru, and
·then turned in under my mosquito curtain, to sleep with
a sense of perfect security in the midst of these good-
natured savages.

We now had seven or eight days of hot and dry
weather, which reduced the little river to a succession of

shallow pools connected by the smallest possible thread of
trickling water. If there were a dry season like that of
Macassar, the Aru Islands would be uninhabitable, as there
is no part of them much above a hundred feet high ; and
the whole being a mass of porous coralline rock, allows
the surface water rapidly to escape. The only dry season
they have is for a month or two about September or
October, and there is then an excessive scarcity of water,
so that sometimes hundreds of birds and other animals die
of drought. The natives then remove to houses near the
sources of the small streams, where, in the shady depths of
the forest, a small quantity of water still remains. Even
then many of them have to go miles for their water, which
they keep in large bamboos and use very sparingly. They
assure me that they catch and kill game of all kinds,
by watching at the water holes or setting snares around
them. That would be the time for me to make my collections ; but the want of water would be a terrible annoyance, and the impossibility of getting away before another
whole year had passed made it out of the question.

Ever since leaving Dobbo I had suffered terribly from
insects, who seemed here bent upon revenging my long-
continued persecution of their race. At our first stopping-
place sand-flies were very abundant at night, penetrating to
every part of the body, and producing a more lasting irritation than mosquitoes. My feet and ankles especially

suffered, and were completely covered with little red swollen specks, which tormented me horribly. On arriving here we were delighted to find the house free from sand-flies or mosquitoes, but in the plantations where my daily walks led me, the day-biting mosquitoes swarmed, and seemed especially to delight in attacking my poor feet. After a month's incessant punishment, those useful members rebelled against such treatment and broke into open insurrection, throwing out numerous inflamed ulcers, which were very painful, and stopped me from walking. So I found myself confined to the house, and with no immediate prospect of leaving it. Wounds or sores in the feet are especially difficult to heal in hot climates, and I therefore dreaded them more than any other illness. The confinement was very annoying, as the fine hot weather was excellent for insects, of which I had every promise of obtaining a fine collection; and it is only by daily and unremitting search that the smaller kinds, and the rarer and more interesting specimens, can be obtained. When I crawled down to the river-side to bathe, I often saw the blue-winged Papilio ulysses, or some other equally rare and beautiful insect; but there was nothing for it but patience, and to return quietly to my bird-skinning, or whatever other work I had indoors. The stings and bites and ceaseless irritation caused by these pests of the tropical forests, would be borne uncomplainingly; but to be kept

prisoner by them in so rich and unexplored a country,
where rare and beautiful creatures are to be met with in
every forest ramble—a country reached by such a long and
tedious voyage, and which might not in the present century be again visited for the same purpose—is a punishment too severe for a naturalist to pass over in silence.

I had, however, some consolation in the birds my boys
brought home daily, more especially the Paradiseas, which
they at length obtained in full plumage. It was quite a
relief to my mind to get these, for I could hardly have torn
myself away from Aru had I not obtained specimens.
But what I valued almost as much as the birds themselves
was the knowledge of their habits, which I was daily obtaining both from the accounts of my hunters, and from
the conversation of the natives. The birds had now commenced what the people here call their "sácaleli," or
dancing-parties, in certain trees in the forest, which are not
fruit trees as I at first imagined, but which have an immense head of spreading branches and large but scattered
leaves, giving a clear space for the birds to play and exhibit
their plumes. On one of these trees a dozen or twenty
full-plumaged male birds assemble together, raise up their
wings, stretch out their necks, and elevate their exquisite
plumes, keeping them in a continual vibration. Between
whiles they fly across from branch to branch in great excitement, so that the whole tree is filled with waving plumes

in every variety of attitude and motion. (See Frontispiece.)
The bird itself is nearly as large as a crow, and is of a rich
coffee brown colour. The head and neck is of a pure straw
yellow above, and rich metallic green beneath. The long
plumy tufts of golden orange feathers spring from the sides
beneath each wing, and when the bird is in repose are
partly concealed by them. At the time of its excitement,
however, the wings are raised vertically over the back, the
head is bent down and stretched out, and the long plumes
are raised up and expanded till they form two magnificent
golden fans, striped with deep red at the base, and fading
off into the pale brown tint of the finely divided and softly
waving points. The whole bird is then overshadowed by
them, the crouching body, yellow head, and emerald green
throat forming but the foundation and setting to the golden
glory which waves above. When seen in this attitude, the
Bird of Paradise really deserves its name, and must be
ranked as one of the most beautiful and most wonderful of
living things. I continued also to get specimens of the
lovely little king-bird occasionally, as well as numbers of
brilliant pigeons, sweet little parroquets, and many curious
small birds, most nearly resembling those of Australia and
New Guinea.

Here, as among most savage people I have dwelt among,
I was delighted with the beauty of the human form—a

beauty of which stay-at-home civilized people can scarcely
have any conception. What are the finest Grecian statues
to the living, moving, breathing men I saw daily around
me ? The unrestrained grace of the naked savage as he
goes about his daily occupations, or lounges at his ease,
must be seen to be understood; and a youth bending his
bow is the perfection of manly beauty. The women,
however, except in extreme youth, are by no means so
pleasant to look at as the men. Their strongly-marked
features are very unfeminine, and hard work, privations,
and very early marriages soon destroy whatever of beauty
or grace they may for a short time possess. Their toilet is
very simple, but also, I am sorry to say, very coarse,
and disgusting. It consists solely of a mat of plaited
strips of palm leaves, worn tight round the body, and
reaching from the hips to the knees. It seems not to be
changed till worn out, is seldom washed, and is generally
very dirty. This is the universal dress, except in a few
cases where Malay "sarongs" have come into use. Their
frizzly hair is tied in a bunch at the back of the head.
They delight in combing, or rather forking it, using for that
purpose a large wooden fork with four diverging prongs,
which answers the purpose of separating and arranging
the long tangled, frizzly mass of cranial vegetation much
better than any comb could do. The only ornaments of
the women are earrings and necklaces, which they arrange

in various tasteful ways. The ends of a necklace are often attached to the earrings, and then looped on to the hair-knot behind. This has really an elegant appearance, the beads hanging gracefully on each side of the head, and by establishing a connexion with the earrings give an appearance of utility to those barbarous ornaments. We recommend this style to the consideration of those of the fair sex who still bore holes in their ears and hang rings thereto. Another style of necklace among these Papuan belles is to wear two, each hanging on one side of the neck and under the opposite arm, so as to cross each other. This has a very pretty appearance, in part due to the contrast of the white beads or kangaroo teeth of which they are composed with the dark glossy skin. The earrings themselves are formed of a bar of copper or silver, twisted so that the ends cross. The men, as usual among savages, adorn themselves more than the women. They wear necklaces, earrings, and finger rings, and delight in a band of plaited grass tight round the arm just below the shoulder, to which they attach a bunch of hair or bright coloured feathers by way of ornament. The teeth of small animals, either alone, or alternately with black or white beads, form their necklaces, and sometimes bracelets also. For these latter, however, they prefer brass wire, or the black, horny, wing-spines of the cassowary, which they consider a charm. Anklets of brass or shell, and tight plaited garters below the knee, complete their ordinary decorations.

Some natives of Kobror from further south, and who are reckoned the worst and least civilized of the Aru tribes, came one day to visit us. They have a rather more than usually savage appearance, owing to the greater amount of ornaments they use—the most conspicuous being a large horseshoe-shaped comb which they wear over the forehead, the ends resting on the temples. The back of the comb is fastened into a piece of wood, which is plated with tin in front, and above is attached a plume of feathers from a cock's tail. In other respects they scarcely differed from the people I was living with. They brought me a couple of birds, some shells and insects, showing that the report of the white man and his doings had reached their country. There was probably hardly a man in Aru who had not by this time heard of me.

Besides the domestic utensils already mentioned, the moveable property of a native is very scanty. He has a good supply of spears and bows and arrows for hunting, a parang, or chopping-knife, and an axe—for the stone age has passed away here, owing to the commercial enterprise of the Bugis and other Malay races. Attached to a belt, or hung across his shoulder, he carries a little skin pouch and an ornamented bamboo, containing betel-nut, tobacco, and lime, and a small German wooden-handled knife is generally stuck between his waist-cloth of bark and his bare skin. Each man also possesses a "cadjan," or sleeping-mat, made of the broad leaves of a pandanus neatly sewn together in three layers. This mat is about four feet square, and when folded has one end sewn up, so that it forms a kind of sack open at one side. In the closed corner the head or feet can be placed, or by carrying it on the head in a shower it forms both coat and umbrella. It doubles up in a small compass for convenient carriage, and then forms a light and elastic cushion, so that on a journey it becomes clothing, house, bedding, and furniture, all in one.

The only ornaments in an Aru house are trophies of the chase—jaws of wild pigs, the heads and backbones of cassowaries, and plumes made from the feathers of the Bird of Paradise, cassowary, and domestic fowl. The spears, shields, knife-handles, and other utensils are more or less carved in fanciful designs, and the mats and leaf boxes are painted or plaited in neat patterns of red, black, and yellow colours. I must not forget these boxes, which are most ingeniously made of the pith of a palm leaf pegged together, lined inside with pandanus leaves, and outside with the same, or with plaited grass. All the joints and angles are covered with strips of split rattan sewn neatly on. The lid is covered with the brown leathery spathe of the Areca palm, which is impervious to water, and the whole box is neat, strong, and well finished. They are made from a few inches to two or three feet long, and being much esteemed by the Malays as clothes-boxes, are a regular article of export from Aru. The natives use the smaller ones for tobacco or betel-nut, but seldom have clothes enough to require the larger ones, which are only made for sale.

Among the domestic animals which may generally be seen in native houses, are gaudy parrots, green, red, and blue, a few domestic fowls, which have baskets hung for them to lay in under the eaves, and who sleep on the ridge, and several half-starved wolfish-looking dogs. Instead of rats and mice there are curious little marsupial animals about the same size, which run about at night and nibble anything eatable that may be left uncovered. Four or five different kinds of ants attack everything not isolated by water, and one kind even swims across that; great spiders lurk in baskets and boxes, or hide in the folds of my mosquito curtain; centipedes and millepedes are found everywhere. I have caught them under my pillow and on my head; while in every box, and under every board which has lain for some days undisturbed, little scorpions are sure to be found snugly ensconced, with their formidable tails quickly turned up ready for attack or defence. Such companions seem very alarming and dangerous, but all combined are not so bad as the irritation of mosquitoes, or of the insect pests often found at home. These latter are a constant and unceasing source of torment and disgust, whereas you may live a long time among scorpions, spiders, and centipedes, ugly and venomous though they are, and get no harm from them. After living twelve years in the tropics, I have never yet been bitten or stung by either.

The lean and hungry dogs before mentioned were my greatest enemies, and kept me constantly on the watch. If my boys left the bird they were skinning for an instant, it was sure to be carried off. Everything eatable had to be hung up to the roof, to be out of their reach. Ali had just finished skinning a fine King Bird of Paradise one day, when he dropped the skin. Before he could stoop to pick it up, one of this famished race had seized upon it, and he only succeeded in rescuing it from its fangs after it was torn to tatters. Two skins of the large Paradisea, which were quite dry and ready to pack away, were incautiously left on my table for the night, wrapped up in paper. The next morning they were gone, and only a few scattered feathers indicated their fate. My hanging shelf was out of their reach; but having stupidly left a box which served as a step, a full-plumaged Paradise bird was next morning missing; and a dog below the house was to be seen still mumbling over the fragments, with the fine golden plumes all trampled in the mud. Every night, as

soon as I was in bed, I could hear them searching about
for what they could devour, under my table, and all about
my boxes and baskets, keeping me in a state of suspense
till morning, lest something of value might incautiously
have been left within their reach.   They would drink the
oil of my floating lamp and eat the wick, and upset or
break my crockery if my lazy boys had neglected to wash
away even the smell of anything eatable.   Bad, however,
as they are here, they were worse in a Dyak's house in
Borneo where I was once staying, for there they gnawed
off the tops of my waterproof boots, ate a large piece out of
an old leather game-bag, besides devouring a portion of my
mosquito curtain !

April 28*th.*—Last evening we had a grand consultation,
which had evidently been arranged and discussed before-
hand.   A number of the natives gathered round me, and
said they wanted to talk.   Two of the best Malay scholars
helped each other, the rest putting in hints and ideas in
their own language.   They told me a long rambling story ;
but, partly owing to their imperfect knowledge of Malay,
partly through my ignorance of local terms, and partly
through the incoherence of their narrative, I could not
make it out very clearly.   It was, however, a tradition,
and I was glad to find they had anything of the kind.   A
long time ago, they said, some strangers came to Aru, and
came here to Wanumbai, and the chief of the Wanumbai

people did not like them, and wanted them to go away,
but they would not go, and so it came to fighting, and
many Aru men were killed, and some, along with the chief,
were taken prisoners, and carried away by the strangers.
Some of the speakers, however, said that he was not carried
away, but went away in his own boat to escape from the
foreigners, and went to the sea and never came back again.
But they all believe that the chief and the people that
went with him still live in some foreign country; and if
they could but find out where, they would send for them
to come back again. Now having some vague idea that
white men must know every country beyond the sea, they
wanted to know if I had met their people in my country
or in the sea. They thought they must be there, for they
could not imagine where else they could be. They had
sought for them everywhere, they said—on the land and in
the sea, in the forest and on the mountains, in the air and
in the sky, and could not find them; therefore, they must
be in my country, and they begged me to tell them, for I
must surely know, as I came from across the great sea. I
tried to explain to them that their friends could not have
reached my country in small boats; and that there were
plenty of islands like Aru all about the sea, which they
would be sure to find. Besides, as it was so long ago, the
chief and all the people must be dead. But they quite
laughed at this idea, and said they were sure they were alive,

for they had proof of it. And then they told me that a good many years ago, when the speakers were boys, some Wokan men who were out fishing met these lost people in the sea, and spoke to them; and the chief gave the Wokan men a hundred fathoms of cloth to bring to the men of Wanumbai, to show that they were alive and would soon come back to them; but the Wokan men were thieves, and kept the cloth, and they only heard of it afterwards; and when they spoke about it, the Wokan men denied it, and pretended they had not received the cloth;—so they were quite sure their friends were at that time alive and somewhere in the sea. And again, not many years ago, a report came to them that some Bugis traders had brought some children of their lost people; so they went to Dobbo to see about it, and the owner of the house, who was now speaking to me, was one who went; but the Bugis man would not let them see the children, and threatened to kill them if they came into his house. He kept the children shut up in a large box, and when he went away he took them with him. And at the end of each of these stories, they begged me in an imploring tone to tell them if I knew where their chief and their people now were.

By dint of questioning, I got some account of the strangers who had taken away their people. They said they were wonderfully strong, and each one could kill a great many Aru men; and when they were wounded, however badly, they spit upon the place, and it immediately became well. And they made a great net of rattans, and entangled their prisoners in it, and sunk them in the water; and the next day, when they pulled the net up on shore, they made the drowned men come to life again, and carried them away.

Much more of the same kind was told me, but in so confused and rambling a manner that I could make nothing out of it, till I inquired how long ago it was that all this happened, when they told me that after their people were taken away the Bugis came in their praus to trade in Aru, and to buy tripang and birds' nests. It is not impossible that something similar to what they related to me really happened when the early Portuguese discoverers first came to Aru, and has formed the foundation for a continually increasing accumulation of legend and fable. I have no doubt that to the next generation, or even before, I myself shall be transformed into a magician or a demigod, a worker of miracles, and a being of supernatural knowledge. They already believe that all the animals I preserve will come to life again; and to their children it will be related that they actually did so. An unusual spell of fine weather setting in just at my arrival has made them believe I can control the seasons; and the simple circumstance of my always walking alone in the forest is a wonder and a mystery to them, as well as

my asking them about birds and animals I have not yet
seen, and showing an acquaintance with their forms,
colours, and habits. These facts are brought against me
when I disclaim knowledge of what they wish me to tell
them. "You must know," say they; "you know everything: you make the fine weather for your men to shoot;
and you know all about our birds and our animals as well
as we do; and you go alone into the forest and are not
afraid." Therefore every confession of ignorance on my
part is thought to be a blind, a mere excuse to avoid telling them too much. My very writing materials and books
are to them weird things; and were I to choose to mystify
them by a few simple experiments with lens and magnet,
miracles without end would in a few years cluster about
me; and future travellers, penetrating to Wanumbai, would
hardly believe that a poor English naturalist, who had resided a few months among them, could have been the
original of the supernatural being to whom so many
marvels were attributed.

For some days I had noticed a good deal of excitement,
and many strangers came and went armed with spears and
cutlasses, bows and shields. I now found there was war
near us—two neighbouring villages having a quarrel about
some matter of local politics that I could not understand.
They told me it was quite a common thing, and that they
are rarely without fighting somewhere near. Individual

quarrels are taken up by villages and tribes, and the non-payment of the stipulated price for a wife is one of the most frequent causes of bitterness and bloodshed. One of the war shields was brought me to look at. It was made of rattans and covered with cotton twist, so as to be both light, strong, and very tough. I should think it would resist any ordinary bullet. About the middle there was an arm-hole with a shutter or flap over it. This enables the arm to be put through and the bow drawn, while the body and face, up to the eyes, remain protected, which cannot be done if the shield is carried on the arm by loops attached at the back in the ordinary way. A few of the young men from our house went to help their friends, but I could not hear that any of them were hurt, or that there was much hard fighting.

May 8th.—I had now been six weeks at Wanumbai, but for more than half the time was laid up in the house with ulcerated feet. My stores being nearly exhausted, and my bird and insect boxes full, and having no immediate prospect of getting the use of my legs again, I determined on returning to Dobbo. Birds had lately become rather scarce, and the Paradise birds had not yet become as plentiful as the natives assured me they would be in another month. The Wanumbai people seemed very sorry at my departure ; and well they might be, for the shells and insects they picked up on the way to and

from their plantations, and the birds the little boys shot
with their bows and arrows, kept them all well supplied
with tobacco and gambir, besides enabling them to accu-
mulate a stock of beads and coppers for future expenses.
The owner of the house was supplied gratis with a little
rice, fish, or salt, whenever he asked for it, which I must
say was not very often.   On parting, I distributed among
them my remnant stock of salt and tobacco, and gave my
host a flask of arrack, and believe that on the whole my
stay with these simple and good-natured people was productive of pleasure and profit to both parties.   I fully
intended to come back; and had I known that circumstances would have prevented my doing so, should have
felt some sorrow in leaving a place where I had first seen
so many rare and beautiful living things, and had so fully
enjoyed the pleasure which fills the heart of the naturalist
when he is so fortunate as to discover a district hitherto
unexplored, and where every day brings forth new and
unexpected treasures.   We loaded our boat in the afternoon, and, starting before daybreak, by the help of a fair
wind reached Dobbo late the same evening.

CHAPTER XXXII.

DOBBO was full to overflowing, and I was obliged to occupy the court-house where the Commissioners hold their sittings. They had now left the island, and I found the situation agreeable, as it was at the end of the village, with a view down the principal street. It was a mere shed, but half of it had a roughly boarded floor, and by putting up a partition and opening a window I made it a very pleasant abode. In one of the boxes I had left in charge of Herr Warzbergen, a colony of small ants had settled and deposited millions of eggs. It was luckily a fine hot day, and by carrying the box some distance from the house, and placing every article in the sunshine for an hour or two, I got rid of them without damage, as they were fortunately a harmless species.

Dobbo now presented an animated appearance. Five or six new houses had been added to the street; the praus were all brought round to the western side of the point, where they were hauled up on the beach, and were being caulked and covered with a thick white lime-plaster for the homeward voyage, making them the brightest and cleanest looking things in the place. Most of the small boats had returned from the " blakang-tana " (back country), as the side of the islands towards New Guinea is called. Piles of firewood were being heaped up behind the houses; sail-makers and carpenters were busy at work; mother-of-pearl shell was being tied up in bundles, and the black and ugly smoked tripang was having a last exposure to the sun before loading. The spare portion of the crews were employed cutting and squaring timber, and boats from Ceram and Goram were constantly unloading their cargoes of sago-cake for the traders' homeward voyage. The fowls, ducks, and goats all looked fat and thriving on the refuse food of a dense population, and the Chinamen's pigs were in a state of obesity that foreboded early death. Parrots and lories and cockatoos, of a dozen different kinds, were suspended on bamboo perches at the doors of the houses, with metallic green or white fruit-pigeons which cooed musically at noon and eventide. Young cassowaries, strangely striped with black and brown, wandered about the houses or gambolled with the playfulness of kittens in the hot sunshine, with sometimes a pretty little kangaroo, caught in the Aru forests, but already tame and graceful as a petted fawn.

Of an evening there were more signs of life than at the time of my former residence. Tom-toms, jews'-harps, and even fiddles were to be heard, and the melancholy Malay songs sounded not unpleasantly far into the night. Almost every day there was a cock-fight in the street. The spectators make a ring, and after the long steel spurs are tied on, and the poor animals are set down to gash and kill each other, the excitement is immense. Those who have made bets scream and yell and jump frantically, if they think they are going to win or lose, but in a very few minutes it is all over; there is a hurrah from the winners, the owners seize their cocks, the winning bird is caressed and admired, the loser is generally dead or very badly wounded, and his master may often be seen plucking out his feathers as he walks away, preparing him for the cooking pot while the poor bird is still alive.

A game at foot-ball, which generally took place at sunset, was, however, much more interesting to me. The ball used is a rather small one, and is made of rattan, hollow, light, and elastic. The player keeps it dancing a little while on his foot, then occasionally on his arm or thigh, till suddenly he gives it a good blow with the hollow of the foot, and sends it flying high in the air. Another player runs to meet it, and at its first bound catches it on his foot and plays in his turn. The ball must never be touched with the hand; but the arm, shoulder, knee, or

thigh are used at pleasure to rest the foot. Two or three
played very skilfully, keeping the ball continually flying
about, but the place was too confined to show off the game
to advantage. One evening a quarrel arose from some
dispute in the game, and there was a great row, and it
was feared there would be a fight about it—not two men
only, but a party of a dozen or twenty on each side, a
regular battle with knives and krisses; but after a large
amount of talk it passed off quietly, and we heard nothing
about it afterwards.

Most Europeans being gifted by nature with a luxuriant
growth of hair upon their faces, think it disfigures them,
and keep up a continual struggle against her by mowing
down every morning the crop which has sprouted up
during the preceding twenty-four hours. Now the men of
Mongolian race are, naturally, just as many of us want to
be. They mostly pass their lives with faces as smooth and
beardless as an infant's. But shaving seems an instinct of
the human race; for many of these people, having no hair
to take off their faces, shave their heads. Others, however, set resolutely to work to force nature to give them a
beard. One of the chief cock-fighters at Dobbo was a
Javanese, a sort of master of the ceremonies of the ring,
who tied on the spurs and acted as backer-up to one of
the combatants. This man had succeeded, by assiduous
cultivation, in raising a pair of moustaches which were a

triumph of art, for they each contained about a dozen
hairs more than three inches long, and which, being well
greased and twisted, were distinctly visible (when not too
far off) as a black thread hanging down on each side of
his mouth. But the beard to match was the difficulty, for
nature had cruelly refused to give him a rudiment of hair
on his chin, and the most talented gardener could not do
much if he had nothing to cultivate. But true genius
triumphs over difficulties. Although there was no hair
proper on the chin, there happened to be, rather on one
side of it, a small mole or freckle which contained (as
such things frequently do) a few stray hairs. These had
been made the most of. They had reached four or five
inches in length, and formed another black thread dangling
from the left angle of the chin. The owner carried this
as if it were something remarkable (as it certainly was) ;
he often felt it affectionately, passed it between his fingers,
and was evidently extremely proud of his moustaches and
beard !

One of the most surprising things connected with Aru
was the excessive cheapness of all articles of European or
native manufacture. We were here two thousand miles
beyond Singapore and Batavia, which are themselves
emporiums of the "far east," in a place unvisited by,
and almost unknown to, European traders ; everything
reached us through at least two or three hands, often

many more; yet English calicoes and American cotton cloths could be bought for 8*s.* the piece, muskets for 15*s.*, common scissors and German knives at three-halfpence each, and other cutlery, cotton goods, and earthenware in the same proportion. The natives of this out-of-the-way country can, in fact, buy all these things at about the same money price as our workmen at home, but in reality very much cheaper, for the produce of a few hours' labour enables the savage to purchase in abundance what are to him luxuries, while to the European they are necessaries of life. The barbarian is no happier and no better off for this cheapness. On the contrary, it has a most injurious effect on him. He wants the stimulus of necessity to force him to labour; and if iron were as dear as silver, and calico as costly as satin, the effect would be beneficial to him. As it is, he has more idle hours, gets a more constant supply of tobacco, and can intoxicate himself with arrack more frequently and more thoroughly; for your Aru man scorns to get half drunk—a tumbler full of arrack is but a slight stimulus, and nothing less than half a gallon of spirit will make him tipsy to his own satisfaction.

It is not agreeable to reflect on this state of things. At least half of the vast multitudes of uncivilized peoples, on whom our gigantic manufacturing system, enormous capital, and intense competition force the produce of our looms and workshops, would be not a whit worse off physically, and would certainly be improved morally, if all the articles with which we supply them were double or treble their present prices. If at the same time the difference of cost, or a large portion of it, could find its way into the pockets of the manufacturing workmen, thousands would be raised from want to comfort, from starvation to health, and would be removed from one of the chief incentives to crime. It is difficult for an Englishman to avoid contemplating with pride our gigantic and ever-increasing manufactures and commerce, and thinking everything good that renders their progress still more rapid, either by lowering the price at which the articles can be produced, or by discovering new markets to which they may be sent. If, however, the question that is so frequently asked of the votaries of the less popular sciences were put here—" Cui bono ? "—it would be found more difficult to answer than had been imagined. The advantages, even to the few who reap them, would be seen to be mostly physical, while the wide-spread moral and intellectual evils resulting from unceasing labour, low wages, crowded dwellings, and monotonous occupations, to perhaps as large a number as those who gain any real advantage, might be held to show a balance of evil so great, as to lead the greatest admirers of our manufactures and commerce to doubt the advisability of their further development. It will be said : " We cannot stop it ;

capital must be employed; our population must be kept
at work; if we hesitate a moment, other nations now hard
pressing us will get ahead, and national ruin will follow."
Some of this is true, some fallacious. It is undoubtedly
a difficult problem which we have to solve; and I am
inclined to think it is this difficulty that makes men conclude that what seems a necessary and unalterable state of
things must be good—that its benefits must be greater
than its evils. This was the feeling of the American
advocates of slavery; they could not see an easy, comfortable way out of it. In our own case, however, it is to be
hoped, that if a fair consideration of the matter in all its
bearings shows that a preponderance of evil arises from
the immensity of our manufactures and commerce—evil
which must go on increasing with their increase—there is
enough both of political wisdom and true philanthropy in
Englishmen, to induce them to turn their superabundant
wealth into other channels. The fact that has led to these
remarks is surely a striking one: that in one of the most
remote corners of the earth savages can buy clothing
cheaper than the people of the country where it is made;
that the weaver's child should shiver in the wintry wind,
unable to purchase articles attainable by the wild natives
of a tropical climate, where clothing is mere ornament or
luxury, should make us pause ere we regard with unmixed
admiration the system which has led to such a result, and

cause us to look with some suspicion on the further extension of that system. It must be remembered too that our
commerce is not a purely natural growth. It has been
ever fostered by the legislature, and forced to an unnatural
luxuriance by the protection of our fleets and armies. The
wisdom and the justice of this policy have been already
doubted. So soon, therefore, as it is seen that the further
extension of our manufactures and commerce would be an
evil, the remedy is not far to seek.

After six weeks' confinement to the house I was at
length well, and could resume my daily walks in the
forest. I did not, however, find it so productive as when
I had first arrived at Dobbo. There was a damp stagnation about the paths, and insects were very scarce. In
some of my best collecting places I now found a mass of
rotting wood, mingled with young shoots, and overgrown
with climbers, yet I always managed to add something daily
to my extensive collections. I one day met with a curious
example of failure of instinct, which, by showing it to be
fallible, renders it very doubtful whether it is anything more
than hereditary habit, dependent on delicate modifications
of sensation. Some sailors cut down a good-sized tree, and,
as is always my practice, I visited it daily for some time in
search of insects. Among other beetles came swarms of
the little cylindrical wood-borers (Platypus, Tesserocerus,

&c.), and commenced making holes in the bark. After a day or two I was surprised to find hundreds of them sticking in the holes they had bored, and on examination discovered that the milky sap of the tree was of the nature of gutta-percha, hardening rapidly on exposure to the air, and glueing the little animals in self-dug graves. The habit of boring holes in trees in which to deposit their eggs, was not accompanied by a sufficient instinctive knowledge of which trees were suitable, and which destructive to them. If, as is very probable, these trees have an attractive odour to certain species of borers, it might very likely lead to their becoming extinct; while other species, to whom the same odour was disagreeable, and who therefore avoided the dangerous trees, would survive, and would be credited by us with an instinct, whereas they would really be guided by a simple sensation.

Those curious little beetles, the Brenthidæ, were very abundant in Aru. The females have a pointed rostrum, with which they bore deep holes in the bark of dead trees, often burying the rostrum up to the eyes, and in these holes deposit their eggs. The males are larger, and have the rostrum dilated at the end, and sometimes terminating in a good-sized pair of jaws. I once saw two males fighting together; each had a fore-leg laid across the neck of the other, and the rostrum bent quite in an attitude of defiance, and looking most ridiculous. Another time, two were fighting for a female, who stood close by busy at her boring. They pushed at each other with their rostra, and clawed and thumped, apparently in the greatest rage, although their coats of mail must have saved both from injury. The small one, however, soon ran away, acknowledging himself vanquished. In most Coleoptera the

MALE BRENTHIDÆ (*Leptorhynchus angustatus*) FIGHTING.

female is larger than the male, and it is therefore interesting, as bearing on the question of sexual selection, that in this case, as in the stag-beetles where the males fight together, they should be not only better armed, but also much larger than the females.

Just as we were going away, a handsome tree, allied to Erythrina, was in blossom, showing its masses of large crimson flowers scattered here and there about the forest.

Could it have been seen from an elevation, it would have had a fine effect; from below I could only catch sight of masses of gorgeous colour in clusters and festoons over-head, about which flocks of blue and orange lories were fluttering and screaming.

A good many people died at Dobbo this season; I believe about twenty. They were buried in a little grove of Casuarinas behind my house. Among the traders was a Mahometan priest, who superintended the funerals, which were very simple. The body was wrapped up in new white cotton cloth, and was carried on a bier to the grave. All the spectators sat down on the ground, and the priest chanted some verses from the Koran. The graves were fenced round with a slight bamboo railing, and a little carved wooden head-post was put to mark the spot. There was also in the village a small mosque, where every Friday the faithful went to pray. This is probably more remote from Mecca than any other mosque in the world, and marks the farthest eastern extension of the Maho-metan religion. The Chinese here, as elsewhere, showed their superior wealth and civilization by tombstones of solid granite brought from Singapore, with deeply-cut inscriptions, the characters of which are painted in red, blue, and gold. No people have more respect for the graves of their relations and friends than this strange, ubiquitous, money-getting people.

Soon after we had returned to Dobbo, my Macassar boy, Baderoon, took his wages and left me, because I scolded him for laziness. He then occupied himself in gambling, and at first had some luck, and bought ornaments, and had plenty of money. Then his luck turned; he lost everything, borrowed money and lost that, and was obliged to become the slave of his creditor till he had worked out the debt. He was a quick and active lad when he pleased, but was apt to be idle, and had such an incorrigible propensity for gambling, that it will very likely lead to his becoming a slave for life.

The end of June was now approaching, the east monsoon had set in steadily, and in another week or two Dobbo would be deserted. Preparations for departure were everywhere visible, and every sunny day (rather rare now) the streets were as crowded and as busy as beehives. Heaps of tripang were finally dried and packed up in sacks; mother-of-pearl shell, tied up with rattans into convenient bundles, was all day long being carried to the beach to be loaded; water-casks were filled, and cloths and mat-sails mended and strengthened for the run home before the strong east wind. Almost every day groups of natives arrived from the most distant parts of the islands, with cargoes of bananas and sugar-cane to exchange for tobacco, sago, bread, and other luxuries, before the general departure. The Chinamen killed their fat pig and made their parting feast, and kindly sent me some pork, and a basin of birds'-nest stew, which had very little more taste than a dish of vermicelli. My boy Ali returned from Wanumbai, where I had sent him alone for a fortnight to buy Paradise birds and prepare the skins; he brought me sixteen glorious specimens, and had he not been very ill with fever and ague might have obtained twice the number. · He had lived with the people whose house I had occupied, and it is a proof of their goodness, if fairly treated, that although he took with him a quantity of silver dollars to pay for the birds they caught, no attempt was made to rob him, which might have been done with the most perfect impunity. He was kindly treated when ill, and was brought back to me with the balance of the dollars he had not spent.

The Wanumbai people, like almost all the inhabitants of the Aru Islands, are perfect savages, and I saw no signs of any religion. There are, however, three or four villages on the coast where schoolmasters from Amboyna reside, and the people are nominally Christians, and are to some extent educated and civilized. I could not get much real knowledge of the customs of the Aru people during the short time I was among them, but they have evidently been considerably influenced by their long association with Mahometan traders. They often bury their dead, although the national custom is to expose the body on a raised stage till it decomposes. Though there is no limit to the number of wives a man may have, they seldom exceed one or two. A wife is regularly purchased from the parents, the price being a large assortment of articles, always including gongs, crockery, and cloth. They told me that some of the tribes kill the old men and women when they can no longer work, but I saw many very old and decrepid people, who seemed pretty well attended to. No doubt all who have much intercourse with the Bugis and Ceramese traders gradually lose many of their native customs, especially as these people often settle in their villages and marry native women.

The trade carried on at Dobbo is very considerable. This year there were fifteen large praus from Macassar, and perhaps a hundred small boats from Ceram, Goram, and Ké. The Macassar cargoes are worth about 1,000*l.* each, and the other boats take away perhaps about 3,000*l.* worth, so that the whole exports may be estimated at 18,000*l.* per annum. The largest and most bulky items are pearl-shell and tripang, or "bêche-de-mer," with smaller quantities of tortoise-shell, edible birds' nests, pearls, ornamental woods, timber, and Birds of Paradise. These are purchased with a variety of goods. Of arrack, about equal in strength to ordinary West India rum, 3,000 boxes, each containing fifteen half-gallon bottles, are consumed annually. Native cloth from Celebes is much esteemed for its durability, and large quantities are sold, as well as white English calico and American unbleached cottons, common crockery, coarse cutlery, muskets, gunpowder, gongs, small brass cannon, and elephants' tusks. These three last articles constitute the wealth of the Aru people, with which they pay for their wives, or which they hoard up as "real property." Tobacco is in immense demand for chewing, and it must be very strong, or an Aru man will not look at it. Knowing how little these people generally work, the mass of produce obtained annually shows that the islands must be pretty thickly inhabited, especially along the coasts, as nine-tenths of the whole are marine productions.

It was on the 2d of July that we left Aru, followed by all the Macassar praus, fifteen in number, who had agreed to sail in company. We passed south of Banda, and then steered due west, not seeing land for three days, till we sighted some low islands west of Bouton. We had a strong and steady south-east wind day and night, which carried us on at about five knots an hour, where a clipper ship would have made twelve. The sky was continually cloudy, dark, and threatening, with occasional drizzling showers, till we were west of Bouru, when it cleared up and we enjoyed the bright sunny skies of the dry season for the rest of our voyage. It is about here, therefore, that the seasons of the eastern and western regions of the

Archipelago are divided. West of this line from June to
December is generally fine, and often very dry, the rest of
the year being the wet season. East of it the weather is
exceedingly uncertain, each island, and each side of an
island, having its own peculiarities. The difference seems
to consist not so much in the distribution of the rainfall
as in that of the clouds and the moistness of the atmosphere. In Aru, for example, when we left, the little
streams were all dried up, although the weather was
gloomy; while in January, February, and March, when we
had the hottest sunshine and the finest days, they were
always flowing. The driest time of all the year in Aru
occurs in September and October, just as it does in Java
and Celebes. The rainy seasons agree, therefore, with
those of the western islands, although the weather is very
different. The Molucca sea is of a very deep blue colour,
quite distinct from the clear light blue of the Atlantic. In
cloudy and dull weather it looks absolutely black, and
when crested with foam has a stern and angry aspect.
The wind continued fair and strong during our whole
voyage, and we reached Macassar in perfect safety on the
evening of the 11th of July, having made the passage
from Aru (more than a thousand miles) in nine and a half
days.

My expedition to the Aru Islands had been eminently
successful. Although I had been for months confined to

the house by illness, and had lost much time by the want
of the means of locomotion, and by missing the right
season at the right place, I brought away with me more
than nine thousand specimens of natural objects, of about
sixteen hundred distinct species. I had made the acquaintance of a strange and little-known race of men; I had
become familiar with the traders of the far East; I had
revelled in the delights of exploring a new fauna and flora,
one of the most remarkable and most beautiful and least-
known in the world; and I had succeeded in the main
object for which I had undertaken the journey—namely,
to obtain fine specimens of the magnificent Birds of Paradise, and to be enabled to observe them in their native
forests. By this success I was stimulated to continue my
researches in the Moluccas and New Guinea for nearly
five years longer, and it is still the portion of my travels to
which I look back with the most complete satisfaction.

CHAPTER XXXIII.

IN this chapter I propose to give a general sketch of the
physical geography of the Aru Islands, and of their
relation to the surrounding countries; and shall thus be
able to incorporate the information obtained from traders,
and from the works of other naturalists, with my own
observations in these exceedingly interesting and little-
known regions.

The Aru group may be said to consist of one very large
central island with a number of small ones scattered round
it. The great island is called by the natives and traders
"Tana-bŭsar" (great or mainland), to distinguish it as a
whole from Dobbo, or any of the detached islands. It is
of an irregular oblong form, about eighty miles from north
to south, and forty or fifty from east to west, in which
direction it is traversed by three narrow channels, dividing
it into four portions. These channels are always called

rivers by the traders, which puzzled me much till I passed
through one of them, and saw how exceedingly applicable
the name was.   The northern channel, called the river of
Watelai, is about a quarter of a mile wide at its entrance,
but soon narrows to about the eighth of a mile, which
width it retains, with little variation, during its whole
length of nearly fifty miles, till it again widens at its
eastern mouth.   Its course is moderately winding, and the
banks are generally dry and somewhat elevated.   In many
places there are low cliffs of hard coralline limestone, more
or less worn by the action of water; while sometimes level
spaces extend from the banks to low ranges of hills a little
inland.   A few small streams enter it from right and left,
at the mouths of which are some little rocky islands.   The
depth is very regular, being from ten to fifteen fathoms,
and it has thus every feature of a true river, but for the salt
water and the absence of a current.   The other two rivers,
whose names are Vorkai and Maykor, are said to be very
similar in general character; but they are rather near
together, and have a number of cross channels intersecting
the flat tract between them.   On the south side of Maykor
the banks are very rocky, and from thence to the southern
extremity of Aru is an uninterrupted extent of rather
elevated and very rocky country, penetrated by numerous
small streams, in the high limestone cliffs bordering which
the edible birds' nests of Aru are chiefly obtained.   All

my informants stated that the two southern rivers are larger than Watelai.

The whole of Aru is low, but by no means so flat as it has been represented, or as it appears from the sea. Most of it is dry rocky ground, with a somewhat undulating surface, rising here and there into abrupt hillocks, or cut into steep and narrow ravines. Except the patches of swamp which are found at the mouths of most of the small rivers, there is no absolutely level ground, although the greatest elevation is probably not more than two hundred feet. The rock which everywhere appears in the ravines and brooks is a coralline limestone, in some places soft and pliable, in others so hard and crystalline as to resemble our mountain limestone.

The small islands which surround the central mass are very numerous; but most of them are on the east side, where they form a fringe, often extending ten or fifteen miles from the main islands. On the west there are very few, Wamma and Pulo Babi being the chief, with Ougia and Wassia at the north-west extremity. On the east side the sea is everywhere shallow, and full of coral; and it is here that the pearl-shells are found which form one of the chief staples of Aru trade. All the islands are covered with a dense and very lofty forest.

The physical features here described are of peculiar interest, and, as far as I am aware, are to some extent unique; for I have been unable to find any other record of an island of the size of Aru crossed by channels which exactly resemble true rivers. How these channels originated were a complete puzzle to me, till, after a long consideration of the whole of the natural phenomena presented by these islands, I arrived at a conclusion which I will now endeavour to explain. There are three ways in which we may conceive islands which are not volcanic to have been formed, or to have been reduced to their present condition, —by elevation, by subsidence, or by separation from a continent or larger island. The existence of coral rock, or of raised beaches far inland, indicates recent elevation; lagoon coral-islands, and such as have barrier or encircling reefs, have suffered subsidence; while our own islands, whose productions are entirely those of the adjacent continent, have been separated from it. Now the Aru Islands are all coral rock, and the adjacent sea is shallow and full of coral; it is therefore evident that they have been elevated from beneath the ocean at a not very distant epoch. But if we suppose that elevation to be the first and only cause of their present condition, we shall find ourselves quite unable to explain the curious river-channels which divide them. Fissures during upheaval would not produce the regular width, the regular depth, or the winding curves which characterise them; and the action of tides and currents during their elevation might form straits of irregular width and depth, but not the river-like channels which actually exist. If, again, we suppose the last movement to have been one of subsidence, reducing the size of the islands, these channels are quite as inexplicable; for subsidence would necessarily lead to the flooding of all low tracts on the banks of the old rivers, and thus obliterate their courses; whereas these remain perfect, and of nearly uniform width from end to end.

Now if these channels have ever been rivers they must have flowed from some higher regions, and this must have been to the east, because on the north and west the sea-bottom sinks down at a short distance from the shore to an unfathomable depth; whereas on the east a shallow sea, nowhere exceeding fifty fathoms, extends quite across to New Guinea, a distance of about a hundred and fifty miles. An elevation of only three hundred feet would convert the whole of this sea into moderately high land, and make the Aru Islands a portion of New Guinea; and the rivers which have their mouths at Utanata and Wamuka, might then have flowed on across Aru, in the channels which are now occupied by salt water. When the intervening land sunk down, we must suppose the land that now constitutes Aru to have remained nearly stationary, a not very improbable supposition, when we consider the great extent of the shallow sea, and the very small amount of depression the land need have undergone to produce it.

But the fact of the Aru Islands having once been connected with New Guinea does not rest on this evidence alone. There is such a striking resemblance between the productions of the two countries as only exists between portions of a common territory. I collected one hundred species of land-birds in the Aru Islands, and about eighty of them have been found on the mainland of New Guinea. Among these are the great wingless cassowary, two species of heavy brush turkeys, and two of short winged thrushes, which could certainly not have passed over the 150 miles of open sea to the coast of New Guinea. This barrier is equally effectual in the case of many other birds which live only in the depths of the forest, as the kinghunters (Dacelo gaudichaudi), the fly-catching wrens (Todopsis), the great crown pigeon (Goura coronata), and the small wood doves (Ptilonopus perlatus, P. aurantiifrons, and P. coronulatus). Now, to show the real effect of such a barrier, let us take the island of Ceram, which is exactly the same distance from New Guinea, but separated from it by a deep sea. Out of about seventy land-birds inhabiting Ceram, only fifteen are found in New Guinea, and none of these are terrestrial or forest-haunting species. The cassowary is distinct; the kingfishers, parrots, pigeons, fly-catchers, honeysuckers, thrushes, and cuckoos, are almost always quite distinct species. More than this, at least twenty genera, which are common to New Guinea and

Aru, do not extend into Ceram, indicating with a force which every naturalist will appreciate, that the two latter countries have received their faunas in a radically different manner.   Again, a true kangaroo is found in Aru, and the same species occurs in Mysol, which is equally Papuan in its productions, while either the same, or one closely allied to it, inhabits New Guinea; but no such animal is found in Ceram, which is only sixty miles from Mysol.   Another small marsupial animal (Perameles doreyanus) is common to Aru and New Guinea.   The insects show exactly the same results.   The butterflies of Aru are all either New Guinea species, or very slightly modified forms; whereas those of Ceram are more distinct than are the birds of the two countries.

It is now generally admitted that we may safely reason on such facts as these, which supply a link in the defective geological record.   The upward and downward movements which any country has undergone, and the succession of such movements, can be determined with much accuracy ; but geology alone can tell us nothing of lands which have entirely disappeared beneath the ocean.   Here physical geography and the distribution of animals and plants are of the greatest service.   By ascertaining the depth of the seas separating one country from another, we can form some judgment of the changes which are taking place.   If there are other evidences of subsidence, a shallow sea

implies a former connexion of the adjacent lands ; but if
this evidence is wanting, or if there is reason to suspect a
rising of the land, then the shallow sea may be the result
of that rising, and may indicate that the two countries will
be joined at some future time, but not that they have
previously been so.   The nature of the animals and plants
inhabiting these countries will, however, almost always
enable us to determine this question.   Mr. Darwin has
shown us how we may determine in almost every case,
whether an island has ever been connected with a continent or larger land, by the presence or absence of terrestrial Mammalia and reptiles.   What he terms " oceanic
islands " possess neither of these groups of animals, though
they may have a luxuriant vegetation, and a fair number of
birds, insects, and land-shells ; and we therefore conclude
that they have originated in mid-ocean, and have never
been connected with the nearest masses of land.   St.
Helena, Madeira, and New Zealand are examples of
oceanic islands.   They possess all other classes of life,
because these have means of dispersion over wide spaces
of sea, which terrestrial mammals and birds have not, as
is fully explained in Sir Charles Lyell's " Principles of
Geology," and Mr. Darwin's " Origin of Species."   On the
other hand, an island may never have been actually connected with the adjacent continents or islands, and yet
may possess representatives of all classes of animals,

because many terrestrial mammals and some reptiles have the means of passing over short distances of sea. But in these cases the number of species that have thus migrated will be very small, and there will be great deficiencies even in birds and flying insects, which we should imagine could easily cross over. The island of Timor (as I have already shown in Chapter XIII.) bears this relation to Australia; for while it contains several birds and insects of Australian forms, no Australian mammal or reptile is found in it, and a great number of the most abundant and characteristic forms of Australian birds and insects are entirely absent. Contrast this with the British Islands, in which a large proportion of the plants, insects, reptiles, and Mammalia of the adjacent parts of the continent are fully represented, while there are no remarkable deficiencies of extensive groups, such as always occur when there is reason to believe there has been no such connexion. The case of Sumatra, Borneo, and Java, and the Asiatic continent is equally clear; many large Mammalia, terrestrial birds, and reptiles being common to all, while a large number more are of closely allied forms. Now, geology has taught us that this representation by allied forms in the same locality implies lapse of time, and we therefore infer that in Great Britain, where almost every species is absolutely identical with those on the Continent, the separation has been very recent; while in Sumatra and

Java, where a considerable number of the continental species are represented by allied forms, the separation was more remote.

From these examples we may see how important a supplement to geological evidence is the study of the geographical distribution of animals and plants, in determining the former condition of the earth's surface; and how impossible it is to understand the former without taking the latter into account. The productions of the Aru Islands offer the strongest evidence that at no very distant epoch they formed a part of New Guinea; and the peculiar physical features which I have described, indicate that they must have stood at very nearly the same level then as they do now, having been separated by the subsidence of the great plain which formerly connected them with it.

Persons who have formed the usual ideas of the vegetation of the tropics—who picture to themselves the abundance and brilliancy of the flowers, and the magnificent appearance of hundreds of forest trees covered with masses of coloured blossoms, will be surprised to hear, that though vegetation in Aru is highly luxuriant and varied, and would afford abundance of fine and curious plants to adorn our hothouses, yet bright and showy flowers are, as a general rule, altogether absent, or so very scarce

as to produce no effect whatever on the general scenery.
To give particulars : I have visited five distinct localities in the islands, I have wandered daily in the forests,
and have passed along upwards of a hundred miles
of coast and river during a period of six months, much
of it very fine weather, and till just as I was about to
leave, I never saw a single plant of striking brilliancy
or beauty, hardly a shrub equal to a hawthorn, or a
climber equal to a honeysuckle ! It cannot be said
that the flowering season had not arrived, for I saw many
herbs, shrubs, and forest trees in flower, but all had
blossoms of a green or greenish-white tint, not superior to
our lime-trees. Here and there on the river banks and
coasts are a few Convolvulaceæ, not equal to our garden
Ipomæas, and in the deepest shades of the forest some
fine scarlet and purple Zingiberaceæ, but so few and
scattered as to be nothing amid the mass of green and
flowerless vegetation. Yet the noble Cycadaceæ and
screw-pines, thirty or forty feet high, the elegant tree ferns,
the lofty palms, and the variety of beautiful and curious
plants which everywhere meet the eye, attest the warmth
and moisture of the tropics, and the fertility of the soil.
It is true that Aru seemed to me exceptionally poor in
flowers, but this is only an exaggeration of a general
tropical feature ; for my whole experience in the equatorial regions of the west and the east has convinced me,

that in the most luxuriant parts of the tropics, flowers are
less abundant, on the average less showy, and are far less
effective in adding colour to the landscape than in temperate climates.   I have never seen in the tropics such brilliant masses of colour as even England can show in her
furze-clad commons, her heathery mountain-sides, her glades
of wild hyacinths, her fields of poppies, her meadows of
buttercups and orchises—carpets of yellow, purple, azure-
blue, and fiery crimson, which the tropics can rarely exhibit.   We have smaller masses of colour in our hawthorn
and crab trees, our holly and mountain-ash, our broom,
foxgloves, primroses, and purple vetches, which clothe
with gay colours the whole length and breadth of our land.
These beauties are all common.   They are characteristic of
the country and the climate; they have not to be sought
for, but they gladden the eye at every step.   In the regions
of the equator, on the other hand, whether it be forest or
savannah, a sombre green clothes universal nature.   You
may journey for hours, and even for days, and meet with
nothing to break the monotony.   Flowers are everywhere
rare, and anything at all striking is only to be met with at
very distant intervals.

The idea that nature exhibits gay colours in the tropics,
and that the general aspect of nature is there more bright
and varied in hue than with us, has even been made the
foundation of theories of art, and we have been forbidden

to use bright colours in our garments, and in the decorations of our dwellings, because it was supposed that we should be thereby acting in opposition to the teachings of nature. The argument itself is a very poor one, since it might with equal justice be maintained, that as we possess faculties for the appreciation of colours, we should make up for the deficiencies of nature and use the gayest tints in those regions where the landscape is most monotonous. But the assumption on which the argument is founded is totally false, so that even if the reasoning were valid, we need not be afraid of outraging nature, by decorating our houses and our persons with all those gay hues which are so lavishly spread over our fields and mountains, our hedges, woods, and meadows.

It is very easy to see what has led to this erroneous view of the nature of tropical vegetation. In our hot-houses and at our flower-shows we gather together the finest flowering plants from the most distant regions of the earth, and exhibit them in a proximity to each other which never occurs in nature. A hundred distinct plants, all with bright, or strange, or gorgeous flowers, make a wonderful show when brought together; but perhaps no two of these plants could ever be seen together in a state of nature, each inhabiting a distant region or a different station. Again, all moderately warm extra-European countries are mixed up with the tropics in general estimation, and a vague idea is formed that whatever is pre-eminently beautiful *must* come from the hottest parts of the earth. But the fact is quite the contrary. Rhododendrons and azaleas are plants of temperate regions, the grandest lilies are from temperate Japan, and a large proportion of our most showy flowering plants are natives of the Himalayas, of the Cape, of the United States, of Chili, or of China and Japan, all temperate regions. True, there are a great number of grand and gorgeous flowers in the tropics, but the proportion they bear to the mass of the vegetation is exceedingly small; so that what appears an anomaly is nevertheless a fact, and the effect of flowers on the general aspect of nature is far less in the equatorial than in the temperate regions of the earth.

CHAPTER XXXIV.

NEW GUINEA.—DOREY.

(MARCH TO JULY 1858.)

AFTER my return from Gilolo to Ternate, in March 1858, I made arrangements for my long-wished-for voyage to the mainland of New Guinea, where I anticipated that my collections would surpass those which I had formed at the Aru Islands. The poverty of Ternate in articles used by Europeans was shown, by my searching in vain through all the stores for such common things as flour, metal spoons, wide-mouthed phials, beeswax, a pen-knife, and a stone or metal pestle and mortar. I took with me four servants : my head man Ali, and a Ternate lad named Jumaat (Friday), to shoot; Lahagi, a steady middle-aged man, to cut timber and assist me in insect-collecting ; and Loisa, a Javanese cook. As I knew I should have to build a house at Dorey, where I was going, I took with me eighty cadjans, or waterproof mats, made of pandanus leaves, to cover over my baggage on first landing, and to help to roof my house afterwards.

We started on the 25th of March in the schooner *Hester Helena*, belonging to my friend Mr. Duivenboden, and bound on a trading voyage along the north coast of New Guinea. Having calms and light airs, we were three days reaching Gané, near the south end of Gilolo, where we stayed to fill up our water-casks and buy a few provisions. We obtained fowls, eggs, sago, plantains, sweet potatoes, yellow pumpkins, chilies, fish, and dried deer's meat; and on the afternoon of the 29th proceeded on our voyage to Dorey harbour. We found it, however, by no means easy to get along; for so near to the equator the monsoons entirely fail of their regularity, and after passing the southern point of Gilolo we had calms, light puffs of wind, and contrary currents, which kept us for five days in sight of the same islands between it and Poppa. A squall then brought us on to the entrance of Dampier's Straits, where we were again becalmed, and were three more days creeping through them. Several native canoes now came off to us from Waigiou on one side, and Batanta on the other, bringing a few common shells, palm-leaf mats, cocoa-nuts, and pumpkins. They were very extravagant in their demands, being accustomed to sell their trifles to whalers and China ships, whose crews will purchase anything at ten times its value. My only purchases were a float belonging to a turtle-spear, carved to resemble a bird, and a very well made palm-leaf box, for which articles I gave a copper ring and a yard of calico. The canoes were very narrow and furnished with an outrigger, and in some of them there was only one man, who seemed to think nothing of coming out alone eight or ten miles from shore. The people were Papuans, much resembling the natives of Aru.

When we had got out of the Straits, and were fairly in the great Pacific Ocean, we had a steady wind for the first time since leaving Ternate, but unfortunately it was dead ahead, and we had to beat against it, tacking on and off the coast of New Guinea. I looked with intense interest on those rugged mountains, retreating ridge behind ridge into the interior, where the foot of civilized man had never trod. There was the country of the cassowary and the tree-kangaroo, and those dark forests produced the most extraordinary and the most beautiful of the feathered inhabitants of the earth—the varied species of Birds of Paradise. A few days more and I hoped to be in pursuit of these, and of the scarcely less beautiful insects which accompany them. We had still, however, for several days only calms and light head-winds, and it was not till the 10th of April that a fine westerly breeze set in, followed by a squally night, which kept us off the entrance of Dorey harbour. The next morning we entered, and came to anchor off the small island of Mansinam, on which dwelt two German missionaries, Messrs. Otto and Geisler. The former immediately came on board to give us welcome, and invited us to go on shore and breakfast with him. We were then introduced to his companion—who was suffering dreadfully from an abscess on the heel, which had confined him to the house for six months—and to his wife, a young German woman, who had been out only three months. Unfortunately she could speak no Malay or English, and had to guess at our compliments on her excellent breakfast by the justice we did to it.

These missionaries were working men, and had been sent out, as being more useful among savages than persons of a higher class. They had been here about two years, and Mr. Otto had already learnt to speak the Papuan language with fluency, and had begun translating some portions of the Bible. The language, however, is so poor that a considerable number of Malay words have to be used; and it is very questionable whether it is possible to convey any idea of such a book, to a people in so low a state of civilization. The only nominal converts yet made are a few of the women; and some few of the children attend school, and are being taught to read, but they make little progress. There is one feature of this mission which I believe will materially interfere with its moral effect. The missionaries are allowed to trade to eke out the very small salaries granted them from Europe, and of course are obliged to carry out the trade principle of buying cheap and selling dear, in order to make a profit. Like all savages the natives are quite careless of the future, and when their small rice crops are gathered they bring a large portion of it to the missionaries, and sell it for knives, beads, axes, tobacco, or any other articles they may require. A few months later, in the wet season, when food is scarce, they come to buy it back again, and give in exchange tortoiseshell, tripang, wild nutmegs, or other produce. Of course the rice is sold at a much higher rate than it was bought, as is perfectly fair and just—and the operation is on the whole thoroughly beneficial to the natives, who would otherwise consume and waste their food when it was abundant, and then starve—yet I cannot imagine that the natives see it in this light. They must look upon the trading missionaries with some suspicion, and cannot feel so sure of their teachings being disinterested, as would be the case if they acted like the Jesuits in Singapore. The first thing to be done by the missionary in attempting to improve savages, is to convince them by his actions that he comes among them for their benefit only, and not for any private ends of his own. To do this he must act in a different way from other men, not trading and taking advantage of the necessities of those who want to sell, but rather giving to those who are in distress. It would be well if he conformed himself in some degree to native customs, and then endeavoured to show how these customs might be gradually modified, so as to be more healthful and more agreeable. A few energetic and devoted men acting in this way might probably effect a decided moral improvement on the lowest savage tribes, whereas trading missionaries, teaching what Jesus said, but not doing as He did, can scarcely be expected to do more than give them a very little of the superficial varnish of religion.

Dorey harbour is in a fine bay, at one extremity of which an elevated point juts out, and, with two or three small islands, forms a sheltered anchorage. The only vessel it contained when we arrived was a Dutch brig, laden with coals for the use of a war-steamer, which was expected daily, on an exploring expedition along the coasts of New Guinea, for the purpose of fixing on a locality for a colony. In the evening we paid it a visit, and landed at the village of Dorey, to look out for a place where I could build my house. Mr. Otto also made arrangements for me with some of the native chiefs, to send men to cut wood, rattans, and bamboo the next day.

The villages of Mansinam and Dorey presented some features quite new to me. The houses all stand completely in the water, and are reached by long rude bridges. They are very low, with the roof shaped like a large boat, bottom upwards. The posts which support the houses, bridges, and platforms are small crooked sticks, placed without any regularity, and looking as if they were tumbling down. The floors are also formed of sticks, equally irregular, and so loose and far apart that I found it almost impossible to walk on them. The walls consist of bits of boards, old boats, rotten mats, attaps, and palm-leaves, stuck in anyhow here and there, and having altogether the most wretched and dilapidated appearance it is possible to conceive. Under the eaves of many of the houses hang human skulls, the trophies of their battles with the savage Arfaks of the interior, who often come to attack them. A large boat-shaped council-house is supported on larger posts, each of which is grossly carved to represent a naked male or female human figure, and other carvings still more revolting are placed upon the platform before the entrance. The view of an ancient lake-dweller's village, given as the frontispiece of Sir Charles Lyell's " Antiquity of Man," is chiefly founded on a sketch of this very village of Dorey; but the extreme regularity of the structures there depicted has no place in the original, any more than it probably had in the actual lake-villages.

The people who inhabit these miserable huts are very similar to the Ké and Aru islanders, and many of them are very handsome, being tall and well-made, with well-cut features and large aquiline noses. Their colour is a deep brown, often approaching closely to black, and the

fine mop-like heads of frizzly hair appear to be more
common than elsewhere, and are considered a great ornament, a long six-pronged bamboo fork being kept stuck in
them to serve the purpose of a comb; and this is assiduously used at idle moments to keep the densely growing

PAPUAN, NEW GUINEA.

mass from becoming matted and tangled. The majority
have short woolly hair, which does not seem capable of
an equally luxuriant development. A growth of hair somewhat similar to this, and almost as abundant, is found

among the half-breeds between the Indian and Negro in
South America. Can this be an indication that the
Papuans are a mixed race?

For the first three days after our arrival I was fully
occupied from morning to night building a house, with the
assistance of a dozen Papuans and my own men. It was
immense trouble to get our labourers to work, as scarcely
one of them could speak a word of Malay; and it was only
by the most energetic gesticulations, and going through a
regular pantomime of what was wanted, that we could get
them to do anything. If we made them understand that a
few more poles were required, which two could have easily
cut, six or eight would insist upon going together, although
we needed their assistance in other things. One morning
ten of them came to work, bringing only one chopper between them, although they knew I had none ready for use.

chose a place about two hundred yards from the beach,
on an elevated ground, by the side of the chief path from
the village of Dorey to the provision-grounds and the forest.
Within twenty yards was a little stream, which furnished
us with excellent water and a nice place to bathe. There
was only low underwood to clear away, while some fine
forest trees stood at a short distance, and we cut down the
wood for about twenty yards round to give us light and
air. The house, about twenty feet by fifteen, was built
entirely of wood, with a bamboo floor, a single door of

thatch, and a large window, looking over the sea, at which I fixed my table, and close beside it my bed, within a little partition. I bought a number of very large palm-leaf mats of the natives, which made excellent walls; while the mats I had brought myself were used on the roof, and were covered over with attaps as soon as we could get them made. Outside, and rather behind, was a little hut, used for cooking, and a bench, roofed over, where my men could sit to skin birds and animals. When all was finished, I had my goods and stores brought up, arranged them conveniently inside, and then paid my Papuans with knives and choppers, and sent them away. The next day our schooner left for the more eastern islands, and I found myself fairly established as the only European inhabitant of the vast island of New Guinea.

As we had some doubt about the natives, we slept at first with loaded guns beside us and a watch set; but after a few days, finding the people friendly, and feeling sure that they would not venture to attack five well-armed men, we took no further precautions. We had still a day or two's work in finishing up the house, stopping leaks, putting up our hanging shelves for drying specimens inside and out, and making the path down to the water, and a clear dry space in front of the house.

On the 17th, the steamer not having arrived, the coal-ship left, having lain here a month, according to her con-

tract; and on the same day my hunters went out to shoot
for the first time, and brought home a magnificent crown
pigeon and a few common birds. The next day they were
more successful, and I was delighted to see them return
with a Bird of Paradise in full plumage, a pair of the fine
Papuan lories (Lorius domicella), four other lories and
parroquets, a grackle (Gracula dumonti), a king-hunter
(Dacelo gaudichaudi), a racquet-tailed kingfisher (Tanysiptera galatea), and two or three other birds of less beauty.
I went myself to visit the native village on the hill behind
Dorey, and took with me a small present of cloth, knives,
and beads, to secure the good-will of the chief, and get
him to send some men to catch or shoot birds for me.
The houses were scattered about among rudely cultivated
clearings. Two which I visited consisted of a central
passage, on each side of which opened short passages, admitting to two rooms, each of which was a house accom-
modating a separate family. They were elevated at least
fifteen feet above the ground, on a complete forest of poles,
and were so rude and dilapidated that some of the small
passages had openings in the floor of loose sticks, through
which a child might fall. The inhabitants seemed rather
uglier than those at Dorey village. They are, no doubt,
the true indigenes of this part of New Guinea, living in
the interior, and subsisting by cultivation and hunting.
The Dorey men, on the other hand, are shore-dwellers,

fishers and traders in a small way, and have thus the character of a colony who have migrated from another district. These hillmen or " Arfaks " differed much in physical features. They were generally black, but some were brown like Malays. Their hair, though always more or less frizzly, was sometimes short and matted, instead of being long, loose, and woolly; and this seemed to be a constitutional difference, not the effect of care and cultivation. Nearly half of them were afflicted with the scurfy skin-disease. The old chief seemed much pleased with

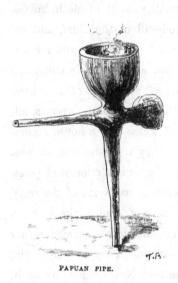

PAPUAN PIPE.

his present, and promised (through an interpreter I brought with me) to protect my men when they came there shooting, and also to procure me some birds and animals. While conversing, they smoked tobacco of their own growing, in pipes cut from a single piece of wood with a long upright handle.

We had arrived at Dorey about the end of the wet season, when the whole country was soaked with moisture. The native paths were so neglected as to be

often mere tunnels closed over with vegetation, and in
such places there was always a fearful accumulation of
mud. To the naked Papuan this is no obstruction. He
wades through it, and the next watercourse makes him
clean again; but to myself, wearing boots and trousers,
it was a most disagreeable thing to have to go up to
my knees in a mud-hole every morning. The man I
brought with me to cut wood fell ill soon after we arrived,
or I would have set him to clear fresh paths in the worst
places. For the first ten days it generally rained every
afternoon and all night; but by going out every hour
of fine weather, I managed to get on tolerably with my
collections of birds and insects, finding most of those
collected by Lesson during his visit in the *Coquille*, as
well as many new ones. It appears, however, that Dorey
is not the place for Birds of Paradise, none of the natives
being accustomed to preserve them. Those sold here are
all brought from Amberbaki, about a hundred miles west,
where the Doreyans go to trade.

The islands in the bay, with the low lands near the
coast, seem to have been formed by recently raised coral
reefs, and are much strewn with masses of coral but little
altered. The ridge behind my house, which runs out
to the point, is also entirely coral rock, although there
are signs of a stratified foundation in the ravines, and
the rock itself is more compact and crystalline. It is

therefore, probably older, a more recent elevation having exposed the low grounds and islands. On the other side of the bay rise the great mass of the Arfak mountains, said by the French navigators to be about ten thousand feet high, and inhabited by savage tribes. These are held in great dread by the Dorey people, who have often been attacked and plundered by them, and have some of their skulls hanging outside their houses. If I was seen going into the forest anywhere in the direction of the mountains, the little boys of the village would shout after me, "Arfaki! Arfaki!" just as they did after Lesson nearly forty years before.

On the 15th of May the Dutch war-steamer *Etna* arrived; but, as the coals had gone, it was obliged to stay till they came back. The captain knew when the coalship was to arrive, and how long it was chartered to stay at Dorey, and could have been back in time, but supposed it would wait for him, and so did not hurry himself. The steamer lay at anchor just opposite my house, and I had the advantage of hearing the half-hourly bells struck, which was very pleasant after the monotonous silence of the forest. The captain, doctor, engineer, and some other of the officers paid me visits; the servants came to the brook to wash clothes, and the son of the Prince of Tidore, with one or two companions, to bathe; otherwise I saw little of them, and was not disturbed by visitors so much as I had expected to be. About this time the weather set in pretty fine, but neither birds nor insects became much more abundant, and new birds were very scarce. None of the Birds of Paradise except the common one were ever met with, and we were still searching in vain for several of the fine birds which Lesson had obtained here. Insects were tolerably abundant, but were not on the average so fine as those of Amboyna, and I reluctantly came to the conclusion that Dorey was not a good collecting locality. Butterflies were very scarce, and were mostly the same as those which I had obtained at Aru.

Among the insects of other orders, the most curious and novel were a group of horned flies, of which I obtained four distinct species, settling on fallen trees and decaying trunks. These remarkable insects, which have been described by Mr. W. W. Saunders as a new genus, under the name of Elaphomia or deer-flies, are about half an inch long, slender-bodied, and with very long legs, which they draw together so as to elevate their bodies high above the surface they are standing upon. The front pair of legs are much shorter, and these are often stretched directly forwards, so as to resemble antennæ. The horns spring from beneath the eye, and seem to be a prolongation of the lower part of the orbit. In the largest and most singular species, named Elaphomia cervicornis or the stag-horned deer-fly, these horns are nearly as long as the body, having two branches, with two small snags near their bifurcation, so as to resemble the horns of a stag. They are black, with the tips pale, while the body and legs are yellowish brown, and the eyes (when alive) violet and green. The next species (Elaphomia wallacei) is of a dark brown

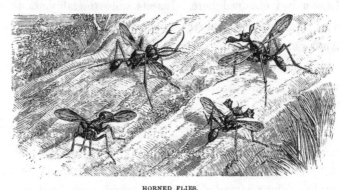

HORNED FLIES.

Elaphomia cervicornis.	Elaphomia wallacei.
E. brevicornis.	E. alcicornis.

colour, banded and spotted with yellow. The horns are about one-third the length of the insect, broad, flat, and of an elongated triangular form. They are of a beautiful pink colour, edged with black, and with a pale central stripe. The front part of the head is also pink, and the eyes violet pink, with a green stripe across them, giving the insect a very elegant and singular appearance. The third species (Elaphomia alcicornis, the elk-horned deer-fly) is a little smaller than the two already described, but

resembling in colour Elaphomia wallacei. The horns are
very remarkable, being suddenly dilated into a flat plate,
strongly toothed round the outer margin, and strikingly
resembling the horns of the elk, after which it has been
named. They are of a yellowish colour, margined with
brown, and tipped with black on the three upper teeth.
The fourth species (Elaphomia brevicornis, the short-
horned deer-fly) differs considerably from the rest. It is
stouter in form, of a nearly black colour, with a yellow
ring at the base of the abdomen; the wings have dusky
stripes, and the head is compressed and dilated laterally,
with very small flat horns, which are black with a pale
centre, and look exactly like the rudiment of the horns of
the two preceding species. None of the females have any
trace of the horns; and Mr. Saunders places in the same
genus a species which has no horns in either sex (Elaphomia polita). It is of a shining black colour, and resembles Elaphomia cervicornis in form, size, and general
appearance. The figures above given represent these
insects of their natural size and in characteristic attitudes.

The natives seldom brought me anything. They are
poor creatures, and rarely shoot a bird, pig, or kangaroo, or
even the sluggish opossum-like Cuscus. The tree-kangaroos
are found here, but must be very scarce, as my hunters,
although out daily in the forest, never once saw them.
Cockatoos, lories, and parroquets were really the only

common birds. Even pigeons were scarce, and in little variety, although we occasionally got the fine crown pigeon, which was always welcome as an addition to our scantily furnished larder.

Just before the steamer arrived I had wounded my ankle by clambering among the trunks and branches of fallen trees (which formed my best hunting grounds for insects), and, as usual with foot wounds in this climate, it turned into an obstinate ulcer, keeping me in the house for several days. When it healed up it was followed by an internal inflammation of the foot, which by the doctor's advice I poulticed incessantly for four or five days, bringing out a severe inflamed swelling on the tendon above the heel. This had to be leeched, and lanced, and doctored with ointments and poultices for several weeks, till I was almost driven to despair,—for the weather was at length fine, and I was tantalized by seeing grand butterflies flying past my door, and thinking of the twenty or thirty new species of insects that I ought to be getting every day. And this, too, in New Guinea!—a country which I might never visit again,—a country which no naturalist had ever resided in before,—a country which contained more strange and new and beautiful natural objects than any other part of the globe. The naturalist will be able to appreciate my feelings, sitting from morning to night in my little hut, unable to move without a crutch, and my only solace the birds my hunters brought in every afternoon, and the few insects caught by my Ternate man, Lahagi, who now went out daily in my place, but who of course did not get a fourth part of what I should have obtained. To add to my troubles all my men were more or less ill, some with fever, others with dysentery or ague; at one time there were three of them besides myself all helpless, the cook alone being well, and having enough to do to wait upon us. The Prince of Tidore and the Resident of Banda were both on board the steamer, and were seeking Birds of Paradise, sending men round in every direction, so that there was no chance of my getting even native skins of the rarer kinds ; and any birds, insects, or animals the Dorey people had to sell were taken on board the steamer, where purchasers were found for everything, and where a larger variety of articles were offered in exchange than I had to show.

After a month's close confinement in the house I was at length able to go out a little, and about the same time I succeeded in getting a boat and six natives to take Ali and Lahagi to Amberbaki, and to bring them back at the end of a month. Ali was charged to buy all the Birds of Paradise he could get, and to shoot and skin all other rare or new birds ; and Lahagi was to collect insects, which I hoped might be more abundant than at Dorey. When I recommenced my daily walks in search of insects, I found a great change in the neighbourhood, and one very agreeable

to me. All the time I had been laid up the ship's crew
and the Javanese soldiers who had been brought in a
tender (a sailing ship which had arrived soon after the
Etna), had been employed cutting down, sawing, and splitting large trees for firewood, to enable the steamer to get
back to Amboyna if the coal-ship did not return; and they
had also cleared a number of wide, straight paths through
the forest in various directions, greatly to the astonishment
of the natives, who could not make out what it all meant.
I had now a variety of walks, and a good deal of dead
wood on which to search for insects ; but notwithstanding
these advantages, they were not nearly so plentiful as I had
found them at Saráwak, or Amboyna, or Batchian, con-
firming my opinion that Dorey was not a good locality.
It is quite probable, however, that at a station a few miles
in the interior, away from the recently elevated coralline
rocks and the influence of the sea air, a much more abundant harvest might be obtained.

One afternoon I went on board the steamer to return
the captain's visit, and was shown some very nice sketches
(by one of the lieutenants), made on the south coast, and
also at the Arfak mountain, to which they had made an
excursion. From these and the captain's description, it
appeared that the people of Arfak were similar to those of
Dorey, and I could hear nothing of the straight-haired race
which Lesson says inhabits the interior, but which no one

has ever seen, and the account of which I suspect has originated in some mistake.   The captain told me he had made
a detailed survey of part of the south coast, and if the coal
arrived should go away at once to Humboldt Bay, in longitude 141° east, which is the line up to which the Dutch
claim New Guinea.   On board the tender I found a
brother naturalist, a German named Rosenberg, who was
draughtsman to the surveying staff.   He had brought two
men with him to shoot and skin birds, and had been able
to purchase a few rare skins from the natives.   Among
these was a pair of the superb Paradise Pie (Astrapia
nigra) in tolerable preservation.   They were brought from
the island of Jobie, which may be its native country, as it
certainly is of the rarer species of crown pigeon (Goura
steursii), one of which was brought alive and sold on board.
Jobie, however, is a very dangerous place, and sailors are
often murdered there when on shore; sometimes the
vessels themselves being attacked.   Wandammen, on the
mainland opposite Jobie, where there are said to be
plenty of birds, is even worse, and at either of these
places my life would not have been worth a week's purchase had I ventured to live alone and unprotected as at
Dorey.   On board the steamer they had a pair of tree-
kangaroos alive.   They differ chiefly from the ground-
kangaroo in having a more hairy tail, not thickened at
the base, and not used as a prop; and by the powerful

claws on the fore-feet, by which they grasp the bark and
branches, and seize the leaves on which they feed. They
move along by short jumps on their hind-feet, which do
not seem particularly well adapted for climbing trees. It
has been supposed that these tree-kangaroos are a special
adaptation to the swampy, half-drowned forests of New
Guinea, in place of the usual form of the group, which is
adapted only to dry ground. Mr. Windsor Earl makes
much of this theory, but, unfortunately for it, the tree-
kangaroos are chiefly found in the northern peninsula of
New Guinea, which is entirely composed of hills and
mountains with very little flat land, while the kangaroo
of the low flat Aru Islands (Dorcopsis asiaticus) is a
ground species. A more probable supposition seems to
be, that the tree-kangaroo has been modified to enable
it to feed on foliage in the vast forests of New Guinea,
as these form the great natural feature which distinguishes that country from Australia.

On June 5th, the coal-ship arrived, having been sent
back from Amboyna, with the addition of some fresh
stores for the steamer. The wood, which had been almost
all taken on board, was now unladen again, the coal taken
in, and on the 17th both steamer and tender left for Humboldt Bay. We were then a little quiet again, and got
something to eat; for while the vessels were here every bit

of fish or vegetable was taken on board, and I had often to make a small parroquet serve for two meals. My men now returned from Amberbaki, but, alas! brought me almost nothing. They had visited several villages, and even went two days' journey into the interior, but could find no skins of Birds of Paradise to purchase, except the common kind, and very few even of those. The birds found were the same as at Dorey, but were still scarcer. None of the natives anywhere near the coast shoot or prepare Birds of Paradise, which come from far in the interior over two or three ranges of mountains, passing by barter from village to village till they reach the sea. There the natives of Dorey buy them, and on their return home sell them to the Bugis or Ternáte traders. It is therefore hopeless for a traveller to go to any particular place on the coast of New Guinea where rare Paradise birds may have been bought, in hopes of obtaining freshly killed specimens from the natives ; and it also shows the scarcity of these birds in any one locality, since from the Amberbaki district, a celebrated place, where at least five or six species have been procured, not one of the rarer ones has been obtained this year. The Prince of Tidore, who would certainly have got them if any were to be had, was obliged to put up with a few of the common yellow ones. I think it probable that a longer residence at Dorey, a little farther in the interior, might show that several of the rarer kinds were found there, as I obtained a single female of the fine scale-breasted Ptiloris magnificus. I was told at Ternate of a bird that is certainly not yet known in Europe, a black King Paradise Bird, with the curled tail and beautiful side plumes of the common species, but all the rest of the plumage glossy black. The people of Dorey knew nothing about this, although they recognised by description most of the other species.

When the steamer left, I was suffering from a severe attack of fever. In about a week I got over this, but it was followed by such a soreness of the whole inside of the mouth, tongue, and gums, that for many days I could put nothing solid between my lips, but was obliged to subsist entirely on slops, although in other respects very well. At the same time two of my men again fell ill, one with fever, the other with dysentery, and both got very bad. I did what I could for them with my small stock of medicines, but they lingered on for some weeks, till on June 26th poor Jumaat died. He was about eighteen years of age, a native, I believe, of Bouton, and a quiet lad, not very active, but doing his work pretty steadily, and as well as he was able. As my men were all Mahometans, I let them bury him in their own fashion, giving them some new cotton cloth for a shroud.

On July 6th the steamer returned from the eastward. The weather was still terribly wet, when, according to rule,

it should have been fine and dry. We had scarcely anything to eat, and were all of us ill. Fevers, colds, and
dysentery were continually attacking us, and made me long
to get away from New Guinea, as much as ever I had
longed to come there. The captain of the *Etna* paid me
a visit, and gave me a very interesting account of his trip.
They had stayed at Humboldt Bay several days, and found
it a much more beautiful and more interesting place than
Dorey, as well as a better harbour. The natives were
quite unsophisticated, being rarely visited except by stray
whalers, and they were superior to the Dorey people,
morally and physically. They went quite naked. Their
houses were some in the water and some inland, and were
all neatly and well built ; their fields were well cultivated,
and the paths to them kept clear and open, in which
respects Dorey is abominable. They were shy at first,
and opposed the boats with hostile demonstrations, bending their bows, and intimating that they would shoot if
an attempt was made to land. Very judiciously the
captain gave way, but threw on shore a few presents, and
after two or three trials they were permitted to land, and
to go about and see the country, and were supplied with
fruits and vegetables. All communication was carried on
with them by signs—the Dorey interpreter, who accompanied the steamer, being unable to understand a word of
their language. No new birds or animals were obtained,

but in their ornaments the feathers of Paradise birds were seen, showing, as might be expected, that these birds range far in this direction, and probably all over New Guinea.

It is curious that a rudimental love of art should co-exist with such a very low state of civilization. The people of Dorey are great carvers and painters. The outsides of the houses, wherever there is a plank, are covered with rude yet characteristic figures. The high-peaked prows of their boats are ornamented with masses of open filagree work, cut out of solid blocks of wood, and often of very tasteful design. As a figure-head, or pinnacle, there is often a human figure, with a head of cassowary feathers to imitate the Papuan "mop." The floats of their fishing-lines, the wooden beaters used in tempering the clay for their pottery, their tobacco-boxes, and other household articles, are covered with carving of tasteful and often elegant design. Did we not already know that such taste and skill are compatible with utter

CARVED TOOL FOR MAKING POTTERY.

barbarism, we could hardly believe that the same people
are, in other matters, utterly wanting in all sense
of order, comfort, or decency.   Yet such is the case.
They live in the most miserable, crazy, and filthy hovels,
which are utterly destitute of anything that can be called
furniture ; not a stool, or bench, or board is seen in them,
no brush seems to be known, and the clothes they wear are
often filthy bark, or rags, or sacking.   Along the paths
where they daily pass to and from their provision grounds,
not an overhanging bough or straggling briar ever seems to
be cut, so that you have to brush through a rank vegetation, creep under fallen trees and spiny creepers, and wade
through pools of mud and mire, which cannot dry up because the sun is not allowed to penetrate.   Their food is
almost wholly roots and vegetables, with fish or game only
as an occasional luxury, and they are consequently very
subject to various skin diseases, the children especially
being often miserable-looking objects, blotched all over
with eruptions and sores.   If these people are not
savages, where shall we find any ?   Yet they have all
a decided love for the fine arts, and spend their
leisure time in executing works whose good taste and
elegance would often be admired in our schools of
design !

During the latter part of my stay in New Guinea the
weather was very wet, my only shooter was ill, and birds

became scarce, so that my only resource was insect-hunt-
ing. I worked very hard every hour of fine weather, and
daily obtained a number of new species. Every dead tree
and fallen log was searched and searched again; and among
the dry and rotting leaves, which still hung on certain
trees which had been cut down, I found an abundant
harvest of minute Coleoptera. Although I never afterwards found so many large and handsome beetles as in
Borneo, yet I obtained here a great variety of species. For
the first two or three weeks, while I was searching out the
best localities, I took about 30 different kinds of beetles a
day, besides about half that number of butterflies, and a
few of the other orders. But afterwards, up to the very
last week, I averaged 49 species a day. On the 31st of
May, I took 78 distinct sorts, a larger number than I had
ever captured before, principally obtained among dead
trees and under rotten bark. A good long walk on a fine
day up the hill, and to the plantations of the natives,
capturing everything not very common that came in my
way, would produce about 60 species; but on the last day
of June I brought home no less than 95 distinct kinds of
beetles, a larger number than I ever obtained in one day
before or since. It was a fine hot day, and I devoted it to
a search among dead leaves, beating foliage, and hunting
under rotten bark, in all the best stations I had discovered
during my walks. I was out from ten in the morning till

three in the afternoon, and it took me six hours' work at
home to pin and set out all the specimens, and to separate
the species. Although I had already been working this
spot daily for two months and a half, and had obtained
over 800 species of Coleoptera, this day's work added 32
new ones. Among these were 4 Longicorns, 2 Carabidæ,
7 Staphylinidæ, 7 Curculionidæ, 2 Copridæ, 4 Chrysomelidæ,
3 Heteromera, 1 Elater, and 1 Buprestis. Even on the
last day I went out, I obtained 16 new species; so that
although I collected over a thousand distinct sorts of
beetles in a space not much exceeding a square mile
during the three months of my residence at Dorey, I
cannot believe that this represents one half the species
really inhabiting the same spot, or a fourth of what might
be obtained in an area extending twenty miles in each
direction.

On the 22d of July the schooner *Hester Helena* arrived,
and five days afterwards we bade adieu to Dorey, without
much regret, for in no place which I have visited have I
encountered more privations and annoyances. Continual
rain, continual sickness, little wholesome food, with a
plague of ants and flies, surpassing anything I had before
met with, required all a naturalist's ardour to encounter;
and when they were uncompensated by great success in
collecting, became all the more insupportable. This long-
thought-of and much-desired voyage to New Guinea had

realized none of my expectations. Instead of being far
better than the Aru Islands, it was in almost everything
much worse. Instead of producing several of the rarer
Paradise birds, I had not even seen one of them, and had
not obtained any one superlatively fine bird or insect.
I cannot deny, however, that Dorey was very rich in
ants. One small black kind was excessively abundant.
Almost every shrub and tree was more or less infested
with it, and its large papery nests were everywhere to
be seen. They immediately took possession of my house,
building a large nest in the roof, and forming papery
tunnels down almost every post. They swarmed on my
table as I was at work setting out my insects, carrying
them off from under my very nose, and even tearing them
from the cards on which they were gummed if I left them
for an instant. They crawled continually over my hands
and face, got into my hair, and roamed at will over my
whole body, not producing much inconvenience till they
began to bite, which they would do on meeting with any
obstruction to their passage, and with a sharpness which
made me jump again and rush to undress and turn out
the offender. They visited my bed also, so that night
brought no relief from their persecutions; and I verily
believe that during my three and a half months' residence
at Dorey I was never for a single hour entirely free from
them. They were not nearly so voracious as many other

kinds, but their numbers and ubiquity rendered it necessary to be constantly on guard against them.

The flies that troubled me most were a large kind of
blue-bottle or blow-fly. These settled in swarms on my
bird skins when first put out to dry, filling their plumage
with masses of eggs, which, if neglected, the next day
produced maggots. They would get under the wings or
under the body where it rested on the drying-board, sometimes actually raising it up half an inch by the mass of
eggs deposited in a few hours; and every egg was so firmly
glued to the fibres of the feathers, as to make it a work of
much time and patience to get them off without injuring
the bird. In no other locality have I ever been troubled
with such a plague as this.

On the 29th we left Dorey, and expected a quick
voyage home, as it was the time of year when we
ought to have had steady southerly and easterly winds.
Instead of these, however, we had calms and westerly
breezes, and it was seventeen days before we reached
Ternate, a distance of five hundred miles only, which,
with average winds, could have been done in five days
It was a great treat to me to find myself back again
in my comfortable house, enjoying milk to my tea and
coffee, fresh bread and butter, and fowl and fish daily
for dinner. This New Guinea voyage had used us all
up, and I determined to stay and recruit before I com-

menced any fresh expeditions. My succeeding journeys
to Gilolo and Batchian have already been narrated, and
it now only remains for me to give an account of my
residence in Waigiou, the last Papuan territory I visited
in search of Birds of Paradise.

CHAPTER XXXV.

VOYAGE FROM CERAM TO WAIGIOU.

(JUNE AND JULY 1860.)

IN my twenty-fifth chapter I have described my arrival at Wahai, on my way to Mysol and Waigiou, islands which belong to the Papuan district, and the account of which naturally follows after that of my visit to the mainland of New Guinea. I now take up my narrative at my departure from Wahai, with the intention of carrying various necessary stores to my assistant, Mr. Allen, at Silinta, in Mysol, and then continuing my journey to Waigiou. It will be remembered that I was travelling in a small prau, which I had purchased and fitted up in Goram, and that, having been deserted by my crew on the coast of Ceram, I had obtained four men at Wahai, who, with my Amboynese hunter, constituted my crew.

Between Ceram and Mysol there are sixty miles of open sea, and along this wide channel the east monsoon blows strongly; so that with native praus, which will not lay up to the wind, it requires some care in crossing. In order to

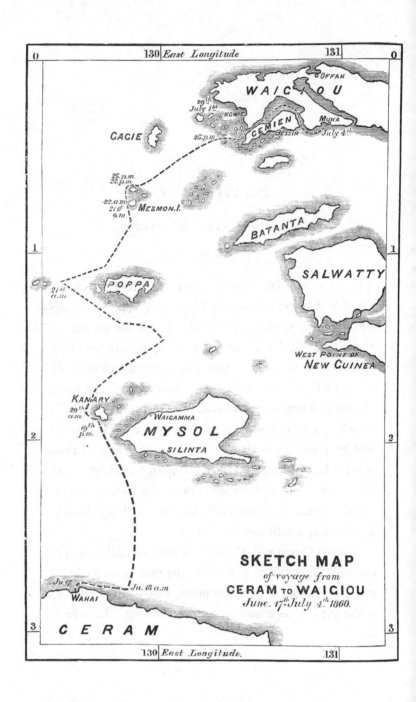

SKETCH MAP
of voyage from
CERAM TO WAICIOU
June. 17th July 4th 1860.

give ourselves sufficient leeway, we sailed back from Wahai eastward, along the coast of Ceram, with the land-breeze; but in the morning (June 18th) had not gone nearly so far as I expected. My pilot, an old and experienced sailor, named Gurulampoko, assured me there was a current setting to the eastward, and that we could easily lay across to Silinta, in Mysol. As we got out from the land the wind increased, and there was a considerable sea, which made my short little vessel plunge and roll about violently. By sunset we had not got halfway across, but could see Mysol distinctly. All night we went along uneasily, and at daybreak, on looking out anxiously, I found that we had fallen much to the westward during the night, owing, no doubt, to the pilot being sleepy and not keeping the boat sufficiently close to the wind. We could see the mountains distinctly, but it was clear we should not reach Silinta, and should have some difficulty in getting to the extreme westward point of the island. The sea was now very boisterous, and our prau was continually beaten to leeward by the waves, and after another weary day we found we could not get to Mysol at all, but might perhaps reach the island called Pulo Kanary, about ten miles to the north-west. Thence we might await a favourable wind to reach Waigamma, on the north side of the island, and visit Allen by means of a small boat.

About nine o'clock at night, greatly to my satisfaction,

we got under the lee of this island, into quite smooth
water—for I had been very sick and uncomfortable, and
had eaten scarcely anything since the preceding morning.
We were slowly nearing the shore, which the smooth dark
water told us we could safely approach, and were congratulating ourselves on soon being at anchor, with the prospect of hot coffee, a good supper, and a sound sleep, when
the wind completely dropped, and we had to get out the
oars to row.   We were not more than two hundred yards
from the shore, when I noticed that we seemed to get no
nearer although the men were rowing hard, but drifted to
the westward; and the prau would not obey the helm, but
continually fell off, and gave us much trouble to bring her
up again.   Soon a loud ripple of water told us we were
seized by one of those treacherous currents which so frequently frustrate all the efforts of the voyager in these
seas; the men threw down the oars in despair, and in a
few minutes we drifted to leeward of the island fairly out
to sea again, and lost our last chance of ever reaching
Mysol!   Hoisting our jib, we lay to, and in the morning
found ourselves only a few miles from the island, but with
such a steady wind blowing from its direction as to render
it impossible for us to get back to it.

We now made sail to the northward, hoping soon to get
a more southerly wind.   Towards noon the sea was much
smoother, and with a S.S.E. wind we were laying in the

direction of Salwatty, which I hoped to reach, as I could there easily get a boat to take provisions and stores to my companion in Mysol. This wind did not, however, last long, but died away into a calm; and a light west wind springing up, with a dark bank of clouds, again gave us hopes of reaching Mysol. We were soon, however, again disappointed. The E.S.E. wind began to blow again with violence, and continued all night in irregular gusts, and with a short cross sea tossed us about unmercifully, and so continually took our sails aback, that we were at length forced to run before it with our jib only, to escape being swamped by our heavy mainsail. After another miserable and anxious night, we found that we had drifted westward of the island of Poppa, and the wind being again a little southerly, we made all sail in order to reach it. This we did not succeed in doing, passing to the north-west, when the wind again blew hard from the E.S.E., and our last hope of finding a refuge till better weather was frustrated. This was a very serious matter to me, as I could not tell how Charles Allen might act, if, after waiting in vain for me, he should return to Wahai, and find that I had left there long before, and had not since been heard of. Such an event as our missing an island forty miles long would hardly occur to him, and he would conclude either that our boat had foundered, or that my crew had murdered me and run away with her. However, as it was physically impossible now for me to reach him, the only thing to be done was to make the best of my way to Waigiou, and trust to our meeting some traders, who might convey to him the news of my safety.

Finding on my map a group of three small islands, twenty-five miles north of Poppa, I resolved, if possible, to rest there a day or two. We could lay our boat's head N.E. by N.; but a heavy sea from the eastward so continually beat us off our course, and we made so much leeway, that I found it would be as much as we could do to reach them. It was a delicate point to keep our head in the best direction, neither so close to the wind as to stop our way, or so free as to carry us too far to leeward. I continually directed the steersman myself, and by incessant vigilance succeeded, just at sunset, in bringing our boat to an anchor under the lee of the southern point of one of the islands. The anchorage was, however, by no means good, there being a fringing coral reef, dry at low water, beyond which, on a bottom strewn with masses of coral, we were obliged to anchor. We had now been incessantly tossing about for four days in our small undecked boat, with constant disappointments and anxiety, and it was a great comfort to have a night of quiet and comparative safety. My old pilot had never left the helm for more than an hour at a time, when one of the others would relieve him for a little sleep; so I determined the next morning to look out for a secure and convenient harbour, and rest on shore for a day.

In the morning, finding it would be necessary for us to get round a rocky point, I wanted my men to go on shore and cut jungle-rope, by which to secure us from being again drifted away, as the wind was directly off shore. I unfortunately, however, allowed myself to be overruled by the pilot and crew, who all declared that it was the easiest thing possible, and that they would row the boat round the point in a few minutes. They accordingly got up the anchor, set the jib, and began rowing; but, just as I had feared, we drifted rapidly off shore, and had to drop anchor again in deeper water, and much farther off. The two best men, a Papuan and a Malay, now swam on shore, each carrying a hatchet, and went into the jungle to seek creepers for rope. After about an hour our anchor loosed hold, and began to drag. This alarmed me greatly, and we let go our spare anchor, and, by running out all our cable, appeared tolerably secure again. We were now most anxious for the return of the men, and were going to fire our muskets to recall them, when we observed them on the beach, some way off, and almost immediately our anchors again slipped, and we drifted slowly away into deep water. We instantly seized the oars, but found we could not counteract the wind and current, and our frantic cries to the men were not heard till we had got a long way off, as they seemed to be hunting for shell-fish on the beach.   Very soon, however, they stared at us, and in a few minutes seemed to comprehend their situation; for they rushed down into the water, as if to swim off, but again returned on shore, as if afraid to make the attempt. We had drawn up our anchors at first not to check our rowing; but now, finding we could do nothing, we let them both hang down by the full length of the cables.   This stopped our way very much, and we drifted from shore very slowly, and hoped the men would hastily form a raft, or cut down a soft-wood tree, and paddle out to us, as we were still not more than a third of a mile from shore. They seemed, however, to have half lost their senses, gesticulating wildly to us, running along the beach, then going into the forest; and just when we thought they had prepared some mode of making an attempt to reach us, we saw the smoke of a fire they had made to cook their shell-fish!   They had evidently given up all idea of coming after us, and we were obliged to look to our own position.

We were now about a mile from shore, and midway between two of the islands, but we were slowly drifting out to sea to the westward, and our only chance of yet saving the men was to reach the opposite shore.   We therefore set our jib and rowed hard; but the wind failed, and we drifted out so rapidly that we had some difficulty in reaching the extreme westerly point of the island. Our only sailor left, then swam ashore with a rope, and helped to tow us round the point into a tolerably safe and secure anchorage, well sheltered from the wind, but exposed to a little swell which jerked our anchor and made us rather uneasy. We were now in a sad plight, having lost our two best men, and being doubtful if we had strength left to hoist our mainsail. We had only two days' water on board, and the small, rocky, volcanic island did not promise us much chance of finding any. The conduct of the men on shore was such as to render it doubtful if they would make any serious attempt to reach us, though they might easily do so, having two good choppers, with which in a day they could make a small outrigger raft on which they could safely cross the two miles of smooth sea with the wind right aft, if they started from the east end of the island, so as to allow for the current. I could only hope they would be sensible enough to make the attempt, and determined to stay as long as I could to give them the chance.

We passed an anxious night, fearful of again breaking our anchor or rattan cable. In the morning (23d), finding all secure, I waded on shore with my two men, leaving the old steersman and the cook on board, with a loaded musket to recall us if needed. We first walked along the beach, till stopped by the vertical cliffs at the east end of the

z 2

island, finding a place where meat had been smoked, a turtle-shell still greasy, and some cut wood, the leaves of which were still green,—showing that some boat had been here very recently. We then entered the jungle, cutting our way up to the top of the hill, but when we got there could see nothing, owing to the thickness of the forest. Returning, we cut some bamboos, and sharpened them to dig for water in a low spot where some sago-trees were growing; when, just as we were going to begin, Hoi, the Wahai man, called out to say he had found water. It was a deep hole among the sago-trees, in stiff black clay, full of water, which was fresh, but smelt horribly from the quantity of dead leaves and sago refuse that had fallen in. Hastily concluding that it was a spring, or that the water had filtered in, we baled it all out as well as a dozen or twenty buckets of mud and rubbish, hoping by night to have a good supply of clean water. I then went on board to breakfast, leaving my two men to make a bamboo raft to carry us on shore and back without wading. I had scarcely finished when our cable broke, and we bumped against the rocks. Luckily it was smooth and calm, and no damage was done. We searched for and got up our anchor, and found that the cable had been cut by grating all night upon the coral. Had it given way in the night, we might have drifted out to sea without our anchor, or been seriously damaged. In the evening we

went to fetch water from the well, when, greatly to our
dismay, we found nothing but a little liquid mud at the
bottom, and it then became evident that the hole was
one which had been made to collect rain water, and would
never fill again as long as the present drought continued.
As we did not know what we might suffer for want of
water, we filled our jar with this muddy stuff so that
it might settle. In the afternoon I crossed over to the
other side of the island, and made a large fire, in order
that our men might see we were still there.

The next day (24th) I determined to have another
search for water; and when the tide was out rounded a
rocky point and went to the extremity of the island
without finding any sign of the smallest stream. On our
way back, noticing a very small dry bed of a watercourse,
I went up it to explore, although everything was so dry
that my men loudly declared it was useless to expect
water there; but a little way up I was rewarded by
finding a few pints in a small pool. We searched higher
up in every hole and channel where water marks appeared,
but could find not a drop more. Sending one of my men
for a large jar and teacup, we searched along the beach till
we found signs of another dry watercourse, and on ascending
this were so fortunate as to discover two deep sheltered
rock-holes containing several gallons of water, enough to
fill all our jars. When the cup came we enjoyed a good

drink of the cool pure water, and before we left had carried away, I believe, every drop on the island.

In the evening a good-sized prau appeared in sight, making apparently for the island where our men were left, and we had some hopes they might be seen and picked up, but it passed along mid-channel, and did not notice the signals we tried to make. I was now, however, pretty easy as to the fate of the men. There was plenty of sago on our rocky island, and there would probably be some on the flat one they were left on. They had choppers, and could cut down a tree and make sago, and would most likely find sufficient water by digging. Shell-fish were abundant, and they would be able to manage very well till some boat should touch there, or till I could send and fetch them. The next day we devoted to cutting wood, filling up our jars with all the water we could find, and making ready to sail in the evening. I shot a small lory closely resembling a common species at Ternate, and a glossy starling which differed from the allied birds of Ceram and Matabello. Large wood-pigeons and crows were the only other birds I saw, but I did not obtain specimens.

About eight in the evening of June 25th we started, and found that with all hands at work we could just haul up our mainsail. We had a fair wind during the night and sailed north-east, finding ourselves in the morning about

twenty miles west of the extremity of Waigiou with a
number of islands intervening. About ten o'clock we ran
full on to a coral reef, which alarmed us a good deal, but
luckily got safe off again. About two in the afternoon we
reached an extensive coral reef, and were sailing close
alongside of it, when the wind suddenly dropped, and
we drifted on to it before we could get in our heavy
mainsail, which we were obliged to let run down and
fall partly overboard. We had much difficulty in getting
off, but at last got into deep water again, though with reefs
and islands all around us. At night we did not know what
to do, as no one on board could tell where we were or what
dangers might surround us, the only one of our crew who
was acquainted with the coast of Waigiou having been
left on the island. We therefore took in all sail and
allowed ourselves to drift, as we were some miles from the
nearest land. A light breeze, however, sprang up, and about
midnight we found ourselves again bumping over a coral reef.
As it was very dark, and we knew nothing of our position,
we could only guess how to get off again, and had there
been a little more wind we might have been knocked to
pieces. However, in about half an hour we did get off,
and then thought it best to anchor on the edge of the
reef till morning. Soon after daylight on the 27th,
finding our prau had received no damage, we sailed on
with uncertain winds and squalls, threading our way

among islands and reefs, and guided only by a small map, which was very incorrect and quite useless, and by a general notion of the direction we ought to take. In the afternoon we found a tolerable anchorage under a small island and stayed for the night, and I shot a large fruit-pigeon new to me, which I have since named Carpophaga tumida. I also saw and shot at the rare white-headed kingfisher (Halcyon saurophaga), but did not kill it. The next morning we sailed on, and having a fair wind reached the shores of the large island of Waigiou. On rounding a point we again ran full on to a coral reef with our mainsail up, but luckily the wind had almost died away, and with a good deal of exertion we managed to get safely off.

We now had to search for the narrow channel among the islands, which we knew was somewhere hereabouts, and which leads to the villages on the south side of Waigiou. Entering a deep bay which looked promising, we got to the end of it, but it was then dusk, so we anchored for the night, and having just finished all our water could cook no rice for supper. Next morning early (29th) we went on shore among the mangroves, and a little way inland found some water, which relieved our anxiety considerably, and left us free to go along the coast in search of the opening, or of some one who could direct us to it. During the three days we had now been among the reefs and islands, we had only seen a single small canoe, which had approached pretty near to us, and then, notwithstanding our signals, went off in another direction. The shores seemed all desert; not a house, or boat, or human being, or a puff of smoke was to be seen; and as we could only go on the course that the ever-changing wind would allow us (our hands being too few to row any distance), our prospects of getting to our destination seemed rather remote and precarious. Having gone to the eastward extremity of the deep bay we had entered, without finding any sign of an opening, we turned westward; and towards evening were so fortunate as to find a small village of seven miserable houses built on piles in the water. Luckily the Orang-kaya, or head man, could speak a little Malay, and informed us that the entrance to the strait was really in the bay we had examined, but that it was not to be seen except when close in-shore. He said the strait was often very narrow, and wound among lakes and rocks and islands, and that it would take two days to reach the large village of Muka, and three more to get to Waigiou. I succeeded in hiring two men to go with us to Muka, bringing a small boat in which to return; but we had to wait a day for our guides, so I took my gun and made a little excursion into the forest. The day was wet and drizzly, and I only succeeded in shooting two small birds, but I saw the great black cockatoo, and had a glimpse of one or two Birds of Paradise, whose loud screams we had heard on first approaching the coast.

Leaving the village the next morning (July 1st) with a light wind, it took us all day to reach the entrance to the channel, which resembled a small river, and was concealed by a projecting point, so that it was no wonder we did not discover it amid the dense forest vegetation which every-where covers these islands to the water's edge. A little way inside it becomes bounded by precipitous rocks, after winding among which for about two miles, we emerged into what seemed a lake, but which was in fact a deep gulf having a narrow entrance on the south coast. This gulf was studded along its shores with numbers of rocky islets, mostly mushroom shaped, from the water having worn away the lower part of the soluble coralline lime-stone, leaving them overhanging from ten to twenty feet. Every islet was covered with strange-looking shrubs and trees, and was generally crowned by lofty and elegant palms, which also studded the ridges of the mountainous shores, forming one of the most singular and picturesque landscapes I have ever seen. The current which had brought us through the narrow strait now ceased, and we were obliged to row, which with our short and heavy prau was slow work. I went on shore several times, but the rocks were so precipitous, sharp, and honeycombed, that I found it impossible to get through the tangled thickets with which they were everywhere clothed. It took us three days to get to the entrance of the gulf, and then the wind was such as to prevent our going any further, and we might have had to wait for days or weeks, when, much to my surprise and gratification, a boat arrived from Muka with one of the head men, who had in some mysterious manner heard I was on my way, and had come to my assistance, bringing a present of cocoa-nuts and vegetables. Being thoroughly acquainted with the coast, and having several extra men to assist us, he managed to get the prau along by rowing, poling, or sailing, and by night had brought us safely into harbour, a great relief after our tedious and unhappy voyage. We had been already eight days among the reefs and islands of Waigiou, coming a distance of about fifty miles, and it was just forty days since we had sailed from Goram.

Immediately on our arrival at Muka, I engaged a small boat and three natives to go in search of my lost men, and sent one of my own men with them to make sure of their going to the right island. In ten days they returned, but to my great regret and disappointment, without the men. The weather had been very bad, and though they had reached an island within sight of that in which the men were, they could get no further. They had waited there six days for better weather, and then, having no more provisions, and the man I had sent with them being very ill and not expected to live, they returned. As they now knew the island, I was determined they should make another trial, and (by a liberal payment of knives, hand-kerchiefs, and tobacco, with plenty of provisions) persuaded them to start back immediately, and make another attempt. They did not return again till the 29th of July, having stayed a few days at their own village of Bessir on the way; but this time they had succeeded and brought with them my two lost men, in tolerable health, though thin and weak. They had lived exactly a month on the island; had found water, and had subsisted on the roots and tender flower-stalks of a species of Bromelia, on shell-fish, and on a few turtles' eggs. Having swum to the island, they had only a pair of trousers and a shirt between them, but had made a hut of palm-leaves, and had altogether got on very well. They saw that I waited for them three days at the opposite island, but had been afraid to cross, lest the current should have carried them out to sea, when they would have been inevitably lost. They had felt sure I would send for them on the first opportunity, and appeared more grateful than natives usually are for my having done so; while I felt much relieved that my voyage, though sufficiently unfortunate, had not involved loss of life.

CHAPTER XXXVI.

WAIGIOU.

(JULY TO SEPTEMBER 1860.)

THE village of Muka, on the south coast of Waigiou, consists of a number of poor huts, partly in the water and partly on shore, and scattered irregularly over a space of about half a mile in a shallow bay. Around it are a few cultivated patches, and a good deal of second-growth woody vegetation; while behind, at the distance of about half a mile, rises the virgin forest, through which are a few paths to some houses and plantations a mile or two inland. The country round is rather flat, and in places swampy, and there are one or two small streams which run behind the village into the sea below it. Finding that no house could be had suitable to my purpose, and having so often experienced the advantages of living close to or just within the forest, I obtained the assistance of half-a-dozen men; and having selected a spot near the path and the stream, and close to a fine fig-tree, which stood just within the forest, we cleared the ground and set to building a house. As I did not expect to stay here so long as I had done at Dorey, I built a long, low, narrow shed, about seven feet high on one side and four on the other, which required but little wood, and was put up very rapidly. Our sails, with a few old attaps from a deserted hut in the village, formed the walls, and a quantity of "cadjans," or palm-leaf mats, covered in the roof. On the third day my house was finished, and all my things put in and comfortably arranged to begin work, and I was quite pleased at having got established so quickly and in such a nice situation.

It had been so far fine weather, but in the night it rained hard, and we found our mat roof would not keep out water. It first began to drop, and then to stream over everything. I had to get up in the middle of the night to secure my insect-boxes, rice, and other perishable articles, and to find a dry place to sleep in, for my bed was soaked. Fresh leaks kept forming as the rain continued, and we all passed a very miserable and sleepless night. In the morning the sun shone brightly, and everything was put out to dry. We tried to find out why the mats leaked, and thought we had discovered that they had been laid on upside down. Having shifted them all, and got everything dry and comfortable by the evening, we again went to bed, and before midnight were again awaked by torrents of rain and leaks streaming in upon us as bad as ever.

There was no more sleep for us that night, and the next day our roof was again taken to pieces, and we came to the conclusion that the fault was a want of slope enough in the roof for mats, although it would be sufficient for the usual attap thatch. I therefore purchased a few new and some old attaps, and in the parts these would not cover we put the mats double, and then at last had the satisfaction of finding our roof tolerably water-tight.

I was now able to begin working at the natural history of the island. When I first arrived I was surprised at being told that there were no Paradise Birds at Muka, although there were plenty at Bessir, a place where the natives caught them and prepared the skins. I assured the people I had heard the cry of these birds close to the village, but they would not believe that I could know their cry. However, the very first time I went into the forest I not only heard but saw them, and was convinced there were plenty about; but they were very shy, and it was some time before we got any. My hunter first shot a female, and I one day got very close to a fine male. He was, as I expected, the rare red species, Paradisea rubra, which alone inhabits this island, and is found nowhere else. He was quite low down, running along a bough searching for insects, almost like a wood-pecker, and the long black riband-like filaments in his tail hung down in the most graceful double curve imagin-

able. I covered him with my gun, and was going to use
the barrel which had a very small charge of powder and
number eight shot, so as not to injure his plumage, but
the gun missed fire, and he was off in an instant among
the thickest jungle. Another day we saw no less than
eight fine males at different times, and fired four times at
them; but though other birds at the same distance almost
always dropped, these all got away, and I began to think
we were not to get this magnificent species. At length the
fruit ripened on the fig-tree close by my house, and many
birds came to feed on it; and one morning, as I was taking
my coffee, a male Paradise Bird was seen to settle on its
top. I seized my gun, ran under the tree, and, gazing up,
could see it flying across from branch to branch, seizing a
fruit here and another there, and then, before I could get
a sufficient aim to shoot at such a height (for it was one of
the loftiest trees of the tropics), it was away into the
forest. They now visited the tree every morning; but
they stayed so short a time, their motions were so rapid,
and it was so difficult to see them, owing to the lower
trees, which impeded the view, that it was only after
several days' watching, and one or two misses, that I
brought down my bird—a male in the most magnificent
plumage.

This bird differs very much from the two large species
which I had already obtained, and, although it wants the

THE RED BIRD OF PARADISE. . (*Paradisea rubra.*)

grace imparted by their long golden trains, is in many respects more remarkable and more beautiful. The head, back, and shoulders are clothed with a richer yellow, the deep metallic green colour of the throat extends further over the head,

and the feathers are elongated on the forehead into two
little erectile crests.   The side plumes are shorter, but are
of a rich red colour, terminating in delicate white points,
and the middle tail-feathers are represented by two long
rigid glossy ribands, which are black, thin, and semi-
cylindrical, and droop gracefully in a spiral curve.   Several
other interesting birds were obtained, and about half-a-
dozen quite new ones; but none of any remarkable beauty,
except the lovely little dove, Ptilonopus pulchellus, which
with several other pigeons I shot on the same fig-tree
close to my house.   It is of a beautiful green colour above,
with a forehead of the richest crimson, while beneath it
is ashy white and rich yellow, banded with violet red.

On the evening of our arrival at Muka I observed what
appeared like a display of Aurora Borealis, though I could
hardly believe that this was possible at a point a little
south of the equator.   The night was clear and calm, and
the northern sky presented a diffused light, with a constant
succession of faint vertical flashings or flickerings, exactly
similar to an ordinary aurora in England.   The next day
was fine, but after that the weather was unprecedentedly
bad, considering that it ought to have been the dry
monsoon.   For near a month we had wet weather; the sun
either not appearing at all, or only for an hour or two
about noon.   Morning and evening, as well as nearly all
night, it rained or drizzled, and boisterous winds, with dark

clouds, formed the daily programme. With the exception
that it was never cold, it was just such weather as a very
bad English November or February.

The people of Waigiou are not truly indigenes of the
island, which possesses no "Alfuros," or aboriginal inhabitants. They appear to be a mixed race, partly from
Gilolo, partly from New Guinea. Malays and Alfuros
from the former island have probably settled here, and
many of them have taken Papuan wives from Salwatty or
Dorey, while the influx of people from those places, and
of slaves, has led to the formation of a tribe exhibiting
almost all the transitions from a nearly pure Malayan to
an entirely Papuan type. The language spoken by them is
entirely Papuan, being that which is used on all the coasts
of Mysol, Salwatty, the north-west of New Guinea, and the
islands in the great Geelvink Bay,—a fact which indicates
the way in which the coast settlements have been formed.
The fact that so many of the islands between New Guinea
and the Moluccas—such as Waigiou, Guebé, Poppa, Obi,
Batchian, as well as the south and east peninsulas of
Gilolo—possess no aboriginal tribes, but are inhabited by
people who are evidently mongrels and wanderers, is a
remarkable corroborative proof of the distinctness of the
Malayan and Papuan races, and the separation of the
geographical areas they inhabit. If these two great races
were direct modifications, the one of the other, we should

expect to find in the intervening region some homogeneous indigenous race presenting intermediate characters. For example, between the whitest inhabitants of Europe and the black Klings of South India, there are in the intervening districts homogeneous races which form a gradual transition from one to the other; while in America, although there is a perfect transition from the Anglo-Saxon to the negro, and from the Spaniard to the Indian, there is no homogeneous race forming a natural transition from one to the other. In the Malay Archipelago we have an excellent example of two absolutely distinct races, which appear to have approached each other, and intermingled in an unoccupied territory at a very recent epoch in the history of man; and I feel satisfied that no unprejudiced person could study them on the spot without being convinced that this is the true solution of the problem, rather than the almost universally accepted view that they are but modifications of one and the same race.

The people of Muka live in that abject state of poverty that is almost always found where the sago-tree is abundant. Very few of them take the trouble to plant any vegetables or fruit, but live almost entirely on sago and fish, selling a little tripang or tortoiseshell to buy the scanty clothing they require. Almost all of them, however, possess one or more Papuan slaves, on whose labour they live in almost absolute idleness, just going out on little fishing or trading excursions, as an excitement in their monotonous existence. They are under the rule of the Sultan of Tidore, and every year have to pay a small tribute of Paradise birds, tortoiseshell, or sago. To obtain these, they go in the fine season on a trading voyage to the mainland of New Guinea, and getting a few goods on credit from some Ceram or Bugis trader, make hard bargains with the natives, and gain enough to pay their tribute, and leave a little profit for themselves.

Such a country is not a very pleasant one to live in, for as there are no superfluities, there is nothing to sell; and had it not been for a trader from Ceram who was residing there during my stay, who had a small vegetable garden, and whose men occasionally got a few spare fish, I should often have had nothing to eat. Fowls, fruit, and vegetables are luxuries very rarely to be purchased at Muka; and even cocoa-nuts, so indispensable for eastern cookery, are not to be obtained; for though there are some hundreds of trees in the village, all the fruit is eaten green, to supply the place of the vegetables the people are too lazy to cultivate. Without eggs, cocoa-nuts, or plantains, we had very short commons, and the boisterous weather being unpropitious for fishing, we had to live on what few eatable birds we could shoot, with an occasional cuscus, or eastern opossum, the only quadruped, except pigs, inhabiting the island.

I had only shot two male Paradiseas on my tree when they ceased visiting it, either owing to the fruit becoming scarce, or that they were wise enough to know there was danger. We continued to hear and see them in the forest, but after a month had not succeeded in shooting any more ; and as my chief object in visiting Waigiou was to get these birds, I determined to go to Bessir, where there are a number of Papuans who catch and preserve them. I hired a small outrigger boat for this journey, and left one of my men to guard my house and goods. We had to wait several days for fine weather, and at length started early one morning, and arrived late at night, after a rough and disagreeable passage. The village of Bessir was built in the water at the point of a small island. The chief food of the people was evidently shell-fish, since great heaps of the shells had accumulated in the shallow water between the houses and the land, forming a regular " kitchen-midden " for the exploration of some future archæologist. We spent the night in the chief's house, and the next morning went over to the mainland to look out for a place where I could reside. This part of Waigiou is really another island to the south of the narrow channel we had passed through in coming to Muka. It appears to consist almost entirely of raised coral, whereas the northern island contains hard crystalline rocks. The shores were a range of low lime- stone cliffs, worn out by the water, so that the upper part

generally overhung. At distant intervals were little coves
and openings, where small streams came down from the
interior; and in one of these we landed, pulling our boat
up on a patch of white sandy beach. Immediately above
was a large newly-made plantation of yams and plantains,
and a small hut, which the chief said we might have the
use of, if it would do for me. It was quite a dwarf's house,
just eight feet square, raised on posts so that the floor was
four and a half feet above the ground, and the highest part
of the ridge only five feet above the floor. As I am six
feet and an inch in my stockings, I looked at this with
some dismay; but finding that the other houses were
much further from water, were dreadfully dirty, and were
crowded with people, I at once accepted the little one, and
determined to make the best of it. At first I thought of
taking out the floor, which would leave it high enough to
walk in and out without stooping; but then there would
not be room enough, so I left it just as it was, had it
thoroughly cleaned out, and brought up my baggage. The
upper story I used for sleeping in, and for a store-room. In
the lower part (which was quite open all round) I fixed up
a small table, arranged my boxes, put up hanging-shelves,
laid a mat on the ground with my wicker-chair upon it,
hung up another mat on the windward side, and then
found that, by bending double and carefully creeping in,
I could sit on my chair with my head just clear of the

ceiling. Here I lived pretty comfortably for six weeks, taking all my meals and doing all my work at my little table, to and from which I had to creep in a semi-horizontal position a dozen times a day; and, after a few severe

MY HOUSE AT BESSIR, IN WAIGIOU.

knocks on the head by suddenly rising from my chair, learnt to accommodate myself to circumstances. We put up a little sloping cooking-hut outside, and a bench on which my lads could skin their birds. At night I went up to my little loft, they spread their mats on the floor below, and we none of us grumbled at our lodgings

My first business was to send for the men who were

accustomed to catch the Birds of Paradise. Several came, and I showed them my hatchets, beads, knives, and hand-kerchiefs; and explained to them, as well as I could by signs, the price I would give for fresh-killed specimens. It is the universal custom to pay for everything in advance; but only one man ventured on this occasion to take goods to the value of two birds. The rest were suspicious, and wanted to see the result of the first bargain with the strange white man, the only one who had ever come to their island. After three days, my man brought me the first bird—a very fine specimen, and alive, but tied up in a small bag, and consequently its tail and wing feathers very much crushed and injured. I tried to explain to him, and to the others that came with him, that I wanted them as perfect as possible, and that they should either kill them, or keep them on a perch with a string to their leg. As they were now apparently satisfied that all was fair, and that I had no ulterior designs upon them, six others took away goods; some for one bird, some for more, and one for as many as six. They said they had to go a long way for them, and that they would come back as soon as they caught any. At intervals of a few days or a week, some of them would return, bringing me one or more birds; but though they did not bring any more in bags, there was not much improvement in their condition. As they caught them a long way off in the forest, they would scarcely

ever come with one, but would tie it by the leg to a
stick, and put it in their house till they caught another.
The poor creature would make violent efforts to escape,
would get among the ashes, or hang suspended by the leg
till the limb was swollen and half-putrefied, and sometimes
die of starvation and worry. One had its beautiful head all
defiled by pitch from a dammar torch; another had been
so long dead that its stomach was turning green. Luckily,
however, the skin and plumage of these birds is so firm
and strong, that they bear washing and cleaning better
than almost any other sort; and I was generally able to
clean them so well that they did not perceptibly differ
from those I had shot myself.

Some few were brought me the same day they were
caught, and I had an opportunity of examining them in
all their beauty and vivacity. As soon as I found they
were generally brought alive, I set one of my men to
make a large bamboo cage with troughs for food and
water, hoping to be able to keep some of them. I got
the natives to bring me branches of a fruit they were
very fond of, and I was pleased to find they ate it
greedily, and would also take any number of live grasshoppers I gave them, stripping off the legs and wings, and
then swallowing them. They drank plenty of water, and
were in constant motion, jumping about the cage from
perch to perch, clinging on the top and sides, and rarely

resting a moment the first day till nightfall. The second day they were always less active, although they would eat as freely as before; and on the morning of the third day they were almost always found dead at the bottom of the cage, without any apparent cause. Some of them ate boiled rice as well as fruit and insects; but after trying many in succession, not one out of ten lived more than three days. The second or third day they would be dull, and in several cases they were seized with convulsions, and fell off the perch, dying a few hours afterwards. I tried immature as well as full-plumaged birds, but with no better success, and at length gave it up as a hopeless task, and confined my attention to preserving specimens in as good a condition as possible.

The Red Birds of Paradise are not shot with blunt arrows, as in the Aru Islands and some parts of New Guinea, but are snared in a very ingenious manner. A large climbing Arum bears a red reticulated fruit, of which the birds are very fond. The hunters fasten this fruit on a stout forked stick, and provide themselves with a fine but strong cord. They then seek out some tree in the forest on which these birds are accustomed to perch, and climbing up it fasten the stick to a branch and arrange the cord in a noose so ingeniously, that when the bird comes to eat the fruit its legs are caught, and by pulling the end of the cord, which hangs down to the ground, it comes free from the branch and brings down the bird. Sometimes, when food is abundant elsewhere, the hunter sits from morning till night under his tree with the cord in his hand, and even for two or three whole days in succession, without even getting a bite; while, on the other hand, if very lucky, he may get two or three birds in a day. There are only eight or ten men at Bessir who practise this art, which is unknown anywhere else in the island. I determined, therefore, to stay as long as possible, as my only chance of getting a good series of specimens; and although I was nearly starved, everything eatable by civilized man being scarce or altogether absent, I finally succeeded.

The vegetables and fruit in the plantations around us did not suffice for the wants of the inhabitants, and were almost always dug up or gathered before they were ripe. It was very rarely we could purchase a little fish; fowls there were none; and we were reduced to live upon tough pigeons and cockatoos, with our rice and sago, and sometimes we could not get these. Having been already eight months on this voyage, my stock of all condiments, spices and butter, was exhausted, and I found it impossible to eat sufficient of my tasteless and unpalatable food to support health. I got very thin and weak, and had a curious disease known (I have since heard) as brow-ague. Directly after breakfast every morning an intense pain set in on a small spot on the right temple. It was a severe

burning ache, as bad as the worst toothache, and lasted
about two hours, generally going off at noon.   When this
finally ceased, I had an attack of fever, which left me so
weak and so unable to eat our regular food, that I feel
sure my life was saved by a couple of tins of soup which
I had long reserved for some such extremity.   I used often
to go out searching after vegetables, and found a great
treasure in a lot of tomato plants run wild, and bearing
little fruits about the size of gooseberries.   I also boiled
up the tops of pumpkin plants and of ferns, by way of
greens, and occasionally got a few green papaws.   The
natives, when hard up for food, live upon a fleshy seaweed, which they boil till it is tender.   I tried this also,
but found it too salt and bitter to be endured.

Towards the end of September it became absolutely
necessary for me to return, in order to make our homeward voyage before the end of the east monsoon.   Most
of the men who had taken payment from me had brought
the birds they had agreed for.   One poor fellow had been
so unfortunate as not to get one, and he very honestly
brought back the axe he had received in advance;
another, who had agreed for six, brought me the fifth
two days before I was to start, and went off immediately
to the forest again to get the other.   He did not return,
however, and we loaded our boat, and were just on the
point of starting, when he came running down after us

holding up a bird, which he handed to me, saying with great satisfaction, " Now I owe you nothing." These were remarkable and quite unexpected instances of honesty among savages, where it would have been very easy for them to have been dishonest without fear of detection or punishment.

The country round about Bessir was very hilly and rugged, bristling with jagged and honey-combed coralline rocks, and with curious little chasms and ravines. The paths often passed through these rocky clefts, which in the depths of the forest were gloomy and dark in the extreme, and often full of fine-leaved herbaceous plants and curious blue-foliaged Lycopodiaceæ. It was in such places as these that I obtained many of my most beautiful small butterflies, such as Sospita statira and Taxila. pulchra, the gorgeous blue Amblypodia hercules, and many others. On the skirts of the plantations I found the handsome blue Deudorix despœna, and in the shady woods the lovely Lycæna wallacei. Here, too, I obtained the beautiful Thyca aruna, of the richest orange on the upper side, while below it is intense crimson and glossy black ; and a superb specimen of a green Ornithoptera, absolutely fresh and perfect, and which still remains one of the glories of my cabinet.

My collection of birds, though not very rich in number

of species, was yet very interesting. I got another specimen of the rare New Guinea kite (Henicopernis longi-
cauda), a large new goatsucker (Podargus superciliaris),
and a most curious ground-pigeon of an entirely new genus,
and remarkable for its long and powerful bill. It has
been named Henicophaps albifrons. I was also much
pleased to obtain a fine series of a large fruit-pigeon with
a protuberance on the bill (Carpophaga tumida), and to
ascertain that this was not, as had been hitherto supposed,
a sexual character, but was found equally in male and
female birds. I collected only seventy-three species of
birds in Waigiou, but twelve of them were entirely new,
and many others very rare; and as I brought away with
me twenty-four fine specimens of the Paradisea rubra, I
did not regret my visit to the island, although it had by
no means answered my expectations.

CHAPTER XXXVII.

VOYAGE FROM WAIGIOU TO TERNATE.

(SEPTEMBER 29 TO NOVEMBER 5, 1860.)

I HAD left the old pilot at Waigiou to take care of my house and to get the prau into sailing order—to caulk her bottom, and to look after the upper works, thatch, and rigging. When I returned I found it nearly ready, and immediately began packing up and preparing for the voyage. Our mainsail had formed one side of our house, but the spanker and jib had been put away in the roof, and on opening them to see if any repairs were wanted, to our horror we found that some rats had made them their nest, and had gnawed through them in twenty places. We had therefore to buy matting and make new sails, and this delayed us till the 29th of September, when we at length left Waigiou.

It took us four days before we could get clear of the land, having to pass along narrow straits beset with reefs and shoals, and full of strong currents, so that an

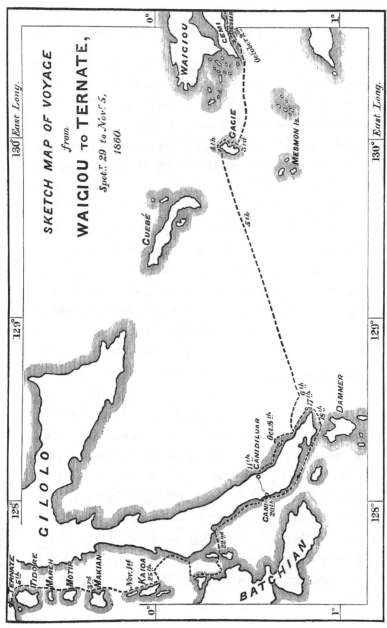

SKETCH MAP OF VOYAGE
from
WAIGIOU TO TERNATE,
Sept.ʳ 29 to Nov.ʳ 5,
1860.

WAIGIOU

CEMI

GAGIE

MESMON Is.

GUEBÉ

CILOLO

CANDILUAR

GANI

DAMMER

BATCHIAN

TERNATE
TIDORE
MAREH
MOTIR
MAKIAN
KAIOA

130° *East Long.*

129°

128°

130° *East Long.*

129°

128°

0°

1°

October 2ⁿᵈ
3ʳᵈ
4ᵗʰ
5ᵗʰ
6ᵗʰ
17ᵗʰ
18ᵗʰ
Oct.8ᵗʰ
11ᵗʰ
20ᵗʰ
23ʳᵈ
25ᵗʰ
Nov.1ˢᵗ
3ʳᵈ
5ᵗʰ

unfavourable wind stopped us altogether.   One day, when
nearly clear, a contrary tide and head wind drove us ten
miles back to our anchorage of the night before.   This
delay made us afraid of running short of water if we
should be becalmed at sea, and we therefore determined,
if possible, to touch at the island where our men had
been lost, and which lay directly in our proper course.
The wind was, however, as usual, contrary, being S.S.W.
instead of S.S.E., as it should have been at this time of the
year, and all we could do was to reach the island of Gagie,
where we came to an anchor by moonlight under bare
volcanic hills.   In the morning we tried to enter a deep
bay, at the head of which some Galela fishermen told us
there was water, but a head-wind preventedus.   For the
reward of a handkerchief, however, they took us to the
place in their boat, and we filled up our jars and bamboos.
We then went round to their camping-place on the north
coast of the island to try and buy something to eat, but
could only get smoked turtle meat as black and as hard
as lumps of coal.   A little further on there was a plantation belonging to Guebe people, but under the care of
a Papuan slave, and the next morning we got some plantains and a few vegetables in exchange for a handkerchief
and some knives.   On leaving this place our anchor had
got foul in some rock or sunken log in very deep water,
and after many unsuccessful attempts, we were forced

to cut our rattan cable and leave it behind us. We had now only one anchor left.

Starting early, on the 4th of October, the same S.S.W. wind continued, and we began to fear that we should hardly clear the southern point of Gilolo. The night of the 5th was squally, with thunder, but after midnight it got tolerably fair, and we were going along with a light wind and looking out for the coast of Gilolo, which we thought we must be nearing, when we heard a dull roaring sound, like a heavy surf, behind us. In a short time the roar increased, and we saw a white line of foam coming on, which rapidly passed us without doing any harm, as our boat rose easily over the wave. At short intervals, ten or a dozen others overtook us with great rapidity, and then the sea became perfectly smooth, as it was before. I concluded at once that these must be earthquake waves ; and on reference to the old voyagers we find that these seas have been long subject to similar phenomena. Dampier encountered them near Mysol and New Guinea, and describes them as follows : " We found here very strange tides, that ran in streams, making a great sea, and roaring so loud that we could hear them before they came within a mile of us. The sea round about them seemed all broken, and tossed the ship so that she would not answer her helm. These ripplings commonly lasted ten or twelve minutes, and then the sea became as still and smooth as a millpond. We

sounded often when in the midst of them, but found no
ground, neither could we perceive that they drove us any
way. We had in one night several of these tides, that
came mostly from the west, and the wind being from that
quarter we commonly heard them a long time before they
came, and sometimes lowered our topsails, thinking it was
a gust of wind. They were of great length, from north to
south, but their breadth not exceeding 200 yards, and they
drove a great pace. For though we had little wind to
move us, yet these would soon pass away, and leave the
water very smooth, and just before we encountered them
we met a great swell, but it did not break." Some time
afterwards, I learnt that an earthquake had been felt on
the coast of Gilolo the very day we had encountered
these curious waves.

When daylight came, we saw the land of Gilolo a few
miles off, but the point was unfortunately a little to windward of us. We tried to brace up all we could to round
it, but as we approached the shore we got into a strong
current setting northward, which carried us so rapidly with
it that we found it necessary to stand off again, in order to
get out of its influence. Sometimes we approached the
point a little, and our hopes revived; then the wind fell,
and we drifted slowly away. Night found us in nearly the
same position as we had occupied in the morning, so we
hung down our anchor with about fifteen fathoms of cable

to prevent drifting.   On the morning of the 7th we were however, a good way up the coast, and we now thought our only chance would be to get close in-shore, where there might be a return current, and we could then row.   The prau was heavy, and my men very poor creatures for work, so that it took us six hours to get to the edge of the reef that fringed the shore; and as the wind might at any moment blow on to it, our situation was a very dangerous one.   Luckily, a short distance off there was a sandy bay, where a small stream stopped the growth of the coral; and by evening we reached this and anchored for the night. Here we found some Galela men shooting deer and pigs; but they could not or would not speak Malay, and we could get little information from them.   We found out that along shore the current changed with the tide, while about a mile out it was always one way, and against us; and this gave us some hopes of getting back to the point, from which we were now distant twenty miles.   Next morning we found that the Galela men had left before daylight, having perhaps some vague fear of our intentions, and very likely taking me for a pirate.   During the morning a boat passed, and the people informed us that, at a short distance further towards the point, there was a much better harbour, where there were plenty of Galela men, from whom we might probably get some assistance.

At three in the afternoon, when the current turned, we

started ; but having a head-wind, made slow progress.
At dusk we reached the entrance of the harbour, but an
eddy and a gust of wind carried us away and out to sea.
After sunset there was a land breeze, and we sailed a little
to the south-east. It then became calm, and we hung
down our anchor forty fathoms, to endeavour to counteract the current; but it was of little avail, and in
the morning we found ourselves a good way from shore,
and just opposite our anchorage of the day before,
which we again reached by hard rowing. I gave the
men this day to rest and sleep ; and the next day
(Oct. 10th) we again started at two in the morning
with a land breeze. After I had set them to their oars,
and given instructions to keep close in-shore, and on
no account to get out to sea, I went below, being rather
unwell. At daybreak I found, to my great astonishment,
that we were again far off-shore, and was told that the
wind had gradually turned more ahead, and had carried
us out—none of them having the sense to take down the
sail and row in-shore, or to call me. As soon as it was
daylight, we saw that we had drifted back, and were again
opposite our former anchorage, and, for the third time, had
to row hard to get to it. As we approached the shore, I
saw that the current was favourable to us, and we continued down the coast till we were close to the entrance
to the lower harbour. Just as we were congratulating

ourselves on having at last reached it, a strong south-east squall came on, blowing us back, and rendering it impossible for us to enter. Not liking the idea of again returning, I determined on trying to anchor, and succeeded in doing so, in very deep water and close to the reefs; but the prevailing winds were such that, should we not hold, we should have no difficulty in getting out to sea. By the time the squall had passed, the current had turned against us, and we expected to have to wait till four in the afternoon, when we intended to enter the harbour.

Now, however, came the climax of our troubles. The swell produced by the squall made us jerk our cable a good deal, and it suddenly snapped low down in the water. We drifted out to sea, and immediately set our mainsail, but we were now without any anchor, and in a vessel so poorly manned that it could not be rowed against the most feeble current or the slightest wind, it would be madness to approach these dangerous shores except in the most perfect calm. We had also only three days' food left. It was therefore out of the question making any further attempts to get round the point without assistance, and I at once determined to run to the village of Gani-diluar, about ten miles further north, where we understood there was a good harbour, and where we might get provisions and a few more rowers. Hitherto winds and currents had invariably opposed our passage southward, and we might have ex-

pected them to be favourable to us now we had turned our
bowsprit in an opposite direction.   But it immediately fell
calm, and then after a time a westerly land breeze set in,
which would not serve us, and we had to row again for
hours, and when night came had not reached the village.
We were so fortunate, however, as to find a deep sheltered
cove where the water was quite smooth, and we constructed a temporary anchor by filling a sack with stones
from our ballast, which being well secured by a network
of rattans held us safely during the night.   The next
morning my men went on shore to cut wood suitable for
making fresh anchors, and about noon, the current turning
in our favour, we proceeded to the village, where we found
an excellent and well-protected anchorage.

On inquiry, we found that the head men resided at the
other Gani on the western side of the peninsula, and it
was necessary to send messengers across (about half a
day's journey) to inform them of my arrival, and to beg
them to assist me.   I then succeeded in buying a little
sago, some dried deer-meat and cocoa-nuts, which at once
relieved our immediate want of something to eat.   At
night we found our bag of stones still held us very well,
and we slept tranquilly.

The next day (October 12th), my men set to work
making anchors and oars.   The native Malay anchor is
ingeniously constructed of a piece of tough forked timber,

the fluke being strengthened by twisted rattans binding
it to the stem, while the cross-piece is formed of a long
flat stone, secured in the same manner. These anchors,
when well made, hold exceedingly firm, and, owing to
the expense of iron, are still almost universally used
on board the smaller praus. In the afternoon the
head men arrived, and promised me as many rowers
as I could put on the prau, and also brought me a few

MALAY ANCHOR.

eggs and a little rice, which were very acceptable. On
the 14th there was a north wind all day, which would
have been invaluable to us a few days earlier, but which
was now only tantalizing. On the 16th, all being ready,
we started at daybreak with two new anchors and ten
rowers, who understood their work. By evening we had
come more than half-way to the point, and anchored for
the night in a small bay. At three the next morning I
ordered the anchor up, but the rattan cable parted close to
the bottom, having been chafed by rocks, and we then lost

our third anchor on this unfortunate voyage. The day was calm, and by noon we passed the southern point of Gilolo, which had delayed us eleven days, whereas the whole voyage during this monsoon should not have occupied more than half that time. Having got round the point our course was exactly in the opposite direction to what it had been, and now, as usual, the wind changed accordingly, coming from the north and north-west,—so that we still had to row every mile up to the village of Gani, which we did not reach till the evening of the 18th. A Bugis trader who was residing there, and the Senaji, or chief, were very kind ; the former assisting me with a spare anchor and a cable, and making me a present of some vegetables, and the latter baking fresh sago cakes for my men, and giving me a couple of fowls, a bottle of oil, and some pumpkins. As the weather was still very uncertain, I got four extra men to accompany me to Ternate, for which place we started on the afternoon of the 20th.

We had to keep rowing all night, the land breezes being too weak to enable us to sail against the current. During the afternoon of the 21st we had an hour's fair wind, which soon changed into a heavy squall with rain, and my clumsy men let the mainsail get taken aback and nearly upset us, tearing the sail, and, what was worse, losing an hour's fair wind. The night was calm, and we made little progress.

On the 22d we had light head-winds. A little before noon we passed, with the assistance of our oars, the Paçiençia Straits, the narrowest part of the channel between Batchian and Gilolo. These were well named by the early Portuguese navigators, as the currents are very strong, and there are so many eddies, that even with a fair wind vessels are often quite unable to pass through them. In the afternoon a strong north wind (dead ahead) obliged us to anchor twice. At night it was calm, and we crept along slowly with our oars.

On the 23d we still had the wind ahead, or calms. We then crossed over again to the mainland of Gilolo by the advice of our Gani men, who knew the coast well. Just as we got across we had another northerly squall with rain, and had to anchor on the edge of a coral reef for the night. I called up my men about three on the morning of the 24th, but there was no wind to help us, and we rowed along slowly. At daybreak there was a fair breeze from the south, but it lasted only an hour. All the rest of the day we had nothing but calms, light winds ahead, and squalls, and made very little progress.

On the 25th we drifted out to the middle of the channel, but made no progress onward. In the afternoon we sailed and rowed to the south end of Kaióa, and by midnight reached the village. I determined to stay here a few days to rest and recruit, and in hopes of getting better weather.

I bought some onions and other vegetables, and plenty of eggs, and my men baked fresh sago cakes. I went daily to my old hunting-ground in search of insects, but with very poor success. It was now wet, squally weather, and there appeared a stagnation of insect life. We stayed five days, during which time twelve persons died in the village, mostly from simple intermittent fever, of the treatment of which the natives are quite ignorant. During the whole of this voyage I had suffered greatly from sun-burnt lips, owing to having exposed myself on deck all day to look after our safety among the shoals and reefs near Waigiou The salt in the air so affected them that they would not heal, but became excessively painful, and bled at the slightest touch, and for a long time it was with great difficulty I could eat at all, being obliged to open my mouth very wide, and put in each mouthful with the greatest caution. I kept them constantly covered with ointment, which was itself very disagreeable, and they caused me almost constant pain for more than a month, as they did not get well till I had returned to Ternate, and was able to remain a week indoors.

A boat which left for Ternate the day after we arrived, was obliged to return the next day, on account of bad weather. On the 31st we went out to the anchorage at the mouth of the harbour, so as to be ready to start at the first favourable opportunity.

On the 1st of November I called up my men at one in the morning, and we started with the tide in our favour. Hitherto it had usually been calm at night, but on this occasion we had a strong westerly squall with rain, which turned our prau broadside, and obliged us to anchor. When it had passed we went on rowing all night, but the wind ahead counteracted the current in our favour, and we advanced but little. Soon after sunrise the wind became stronger and more adverse, and as we had a dangerous lee-shore which we could not clear, we had to put about and get an offing to the W.S.W. This series of contrary winds and bad weather ever since we started, not having had a single day of fair wind, was very remarkable. My men firmly believed there was something unlucky in the boat, and told me I ought to have had a certain ceremony gone through before starting, consisting of boring a hole in the bottom and pouring some kind of holy oil through it. It must be remembered that this was the season of the south-east monsoon, and yet we had not had even half a day's south-east wind since we left Waigiou. Contrary winds, squalls, and currents drifted us about the rest of the day at their pleasure. The night was equally squally and changeable, and kept us hard at work taking in and making sail, and rowing in the intervals.

Sunrise on the 2d found us in the middle of the ten-mile channel between Kaióa and Makian. Squalls and showers succeeded each other during the morning. At noon there was a dead calm, after which a light westerly breeze enabled us to reach a village on Makian in the evening. Here I bought some pumelos (Citrus decumana), kanary-nuts, and coffee, and let my men have a night's sleep.

The morning of the 3d was fine, and we rowed slowly along the coast of Makian. The captain of a small prau at anchor, seeing me on deck and guessing who I was, made signals for us to stop, and brought me a letter from Charles Allen, who informed me he had been at Ternate twenty days, and was anxiously waiting my arrival. This was good news, as I was equally anxious about him, and it cheered up my spirits. A light southerly wind now sprung up, and we thought we were going to have fine weather. It soon changed, however, to its old quarter, the west; dense clouds gathered over the sky, and in less than half an hour we had the severest squall we had experienced during our whole voyage. Luckily we got our great main-sail down in time, or the consequences might have been serious. It was a regular little hurricane, and my old Bugis steersman began shouting out to "Allah! il Allah!" to preserve us. We could only keep up our jib, which was almost blown to rags, but by careful handling it kept us before the wind, and the prau behaved very well. Our small boat (purchased at Gani) was towing astern, and soon got full of water, so that it broke away and we saw no more of it. In about an hour the fury of the wind abated a little, and in two more we were able to hoist our mainsail, reefed and half-mast high. Towards evening it cleared up and fell calm, and the sea, which had been rather high, soon went down. Not being much of a seaman myself I had been considerably alarmed, and even the old steersman assured me he had never been in a worse squall all his life. He was now more than ever confirmed in his opinion of the unluckiness of the boat, and in the efficiency of the holy oil which all Bugis praus had poured through their bottoms. As it was, he imputed our safety and the quick termination of the squall entirely to his own prayers, saying with a laugh, " Yes, that's the way we always do on board our praus ; when things are at the worst we stand up and shout out our prayers as loud as we can, and then Tuwan Allah helps us."

After this it took us two days more to reach Ternate, having our usual calms, squalls, and head-winds to the very last; and once having to return back to our anchorage owing to violent gusts of wind just as we were close to the town. Looking at my whole voyage in this vessel from the time when I left Goram in May, it will appear that my experiences of travel in a native prau have not been encouraging. My first crew ran away; two men were lost for a month on a desert island; we were ten times aground on coral reefs; we lost four anchors; the sails were devoured by rats; the small boat was lost astern; we were thirty-eight days on the voyage home, which should not have taken twelve; we were many times short of food and water; we had no compass-lamp, owing to there not being a drop of oil in Waigiou when we left; and to crown all, during the whole of our voyages from Goram by Ceram to Waigiou, and from Waigiou to Ternate, occupying in all seventy-eight days, or only twelve days short of three months (all in what was supposed to be the favourable season), we had *not one single day of fair wind.* We were always close braced up, always struggling against wind, tide, and leeway, and in a vessel that would scarcely sail nearer than eight points from the wind. Every seaman will admit that my first voyage in my own boat was a most unlucky one.

Charles Allen had obtained a tolerable collection of birds and insects at Mysol, but far less than he would have done if I had not been so unfortunate as to miss visiting him. After waiting another week or two till he was nearly starved, he returned to Wahai in Ceram, and heard, much to his surprise, that I had left a fortnight before. He was delayed there more than a month before he could get back to the north side of Mysol, which he found a much better locality, but it was not yet the season for the Paradise Birds; and before he had obtained more than a few of the common sort, the last prau was ready to leave for Ternate, and he was obliged to take the opportunity, as he expected I would be waiting there for him.

This concludes the record of my wanderings. I next went to Timor, and afterwards to Bouru, Java, and Sumatra, which places have already been described. Charles Allen made a voyage to New Guinea, a short account of which will be given in my next chapter on the Birds of Paradise. On his return he went to the Sula Islands, and made a very interesting collection which served to determine the limits of the zoological group of Celebes, as already explained in my chapter on the natural history of that island. His next journey was to Flores and Solor, where he obtained some valuable materials, which I have used in my chapter on the natural history of the Timor group. He afterwards went to Coti on the east coast of Borneo, from which place I was very anxious to obtain collections, as it is a quite new locality as far as possible from Saráwak, and I had heard very good accounts of it. On his return thence to Sourabaya in Java, he was to have gone to the entirely unknown Sumba or Sandal-wood Island. Most unfortunately, however, he was seized with a terrible fever on his arrival at Coti, and, after lying there some weeks, was taken to

Singapore in a very bad condition, where he arrived after I had left for England. When he recovered he obtained employment in Singapore, and I lost his services as a collector.

The three concluding chapters of my work will treat of the Birds of Paradise, the Natural History of the Papuan Islands, and the Races of Man in the Malay Archipelago.

THE "KING" AND THE "TWELVE WIRED" BIRDS OF PARADISE.

CHAPTER XXXVIII.

THE BIRDS OF PARADISE.

AS many of my journeys were made with the express object of obtaining specimens of the Birds of Paradise, and learning something of their habits and distribution ; and being (as far as I am aware) the only Englishman who has seen these wonderful birds in their native forests, and obtained specimens of many of them, I propose to give here, in a connected form, the result of my observations and inquiries.

When the earliest European voyagers reached the Moluccas in search of cloves and nutmegs, which were then rare and precious spices, they were presented with the dried skins of birds so strange and beautiful as to excite the admiration even of those wealth-seeking rovers. The Malay traders gave them the name of "Manuk dewata," or God's birds ; and the Portuguese, finding that they had no feet or wings, and not being able to learn anything authentic about them, called them " Passaros de Sol," or

Birds of the Sun ; while the learned Dutchmen, who wrote in Latin, called them " Avis paradiseus," or Paradise Bird. John van Linschoten gives these names in 1598, and tells us that no one has seen these birds alive, for they live in the air, always turning towards the sun, and never lighting on the earth till they die ; for they have neither feet nor wings, as, he adds, may be seen by the birds carried to India, and sometimes to Holland, but being very costly they were then rarely seen in Europe. More than a hundred years later Mr. William Funnel, who accompanied Dampier, and wrote an account of the voyage, saw specimens at Amboyna, and was told that they came to Banda to eat nutmegs, which intoxicated them and made them fall down senseless, when they were killed by ants. Down to 1760, when Linnæus named the largest species, Paradisea apoda (the footless Paradise Bird), no perfect specimen had been seen in Europe, and absolutely nothing was known about them. And even now, a hundred years later, most books state that they migrate annually to Ternate, Banda, and Amboyna ; whereas the fact is, that they are as completely unknown in those islands in a wild state as they are in England. Linnæus was also acquainted with a small species, which he named Paradisea regia (the King Bird of Paradise), and since then nine or ten others have been named, all of which were first described from skins preserved by the savages of

New Guinea, and generally more or less imperfect. These are now all known in the Malay Archipelago as "Burong mati," or dead birds, indicating that the Malay traders never saw them alive.

The Paradiseidæ are a group of moderate-sized birds, allied in their structure and habits to crows, starlings, and to the Australian honeysuckers; but they are characterised by extraordinary developments of plumage, which are unequalled in any other family of birds. In several species large tufts of delicate bright-coloured feathers spring from each side of the body beneath the wings, forming trains, or fans, or shields; and the middle feathers of the tail are often elongated into wires, twisted into fantastic shapes, or adorned with the most brilliant metallic tints. In another set of species these accessory plumes spring from the head, the back, or the shoulders; while the intensity of colour and of metallic lustre displayed by their plumage, is not to be equalled by any other birds, except, perhaps, the humming-birds, and is not surpassed even by these. They have been usually classified under two distinct families, Paradiseidæ and Epimachidæ, the latter characterised by long and slender beaks, and supposed to be allied to the Hoopoes; but the two groups are so closely allied in every essential point of structure and habits, that I shall consider them as forming subdivisions of one family. I will now give a short descrip-

tion of each of the known species, and then add some
general remarks on their natural history.

The Great Bird of Paradise (Paradisea apoda of Linnæus) is the largest species known, being generally
seventeen or eighteen inches from the beak to the tip of
the tail. The body, wings, and tail are of a rich coffee-
brown, which deepens on the breast to a blackish-violet
or purple-brown. The whole top of the head and neck
is of an exceedingly delicate straw-yellow, the feathers
being short and close set, so as to resemble plush or
velvet; the lower part of the throat up to the eye is
clothed with scaly feathers of an emerald green colour,
and with a rich metallic gloss, and velvety plumes of a
still deeper green extend in a band across the forehead
and chin as far as the eye, which is bright yellow. The
beak is pale lead blue; and the feet, which are rather
large and very strong and well formed, are of a pale ashy-
pink. The two middle feathers of the tail have no webs,
except a very small one at the base and at the extreme tip,
forming wire-like cirrhi, which spread out in an elegant
double curve, and vary from twenty-four to thirty-four
inches long. From each side of the body, beneath the wings,
springs a dense tuft of long and delicate plumes, sometimes
two feet in length, of the most intense golden-orange colour
and very glossy, but changing towards the tips into a pale

brown. This tuft of plumage can be elevated and spread out at pleasure, so as almost to conceal the body of the bird.

These splendid ornaments are entirely confined to the male sex, while the female is really a very plain and ordinary-looking bird of a uniform coffee-brown colour which never changes, neither does she possess the long tail wires, nor a single yellow or green feather about the head. The young males of the first year exactly resemble the females, so that they can only be distinguished by dissection. The first change is the acquisition of the yellow and green colour on the head and throat, and at the same time the two middle tail feathers grow a few inches longer than the rest, but remain webbed on both sides. At a later period these feathers are replaced by the long bare shafts of the full length, as in the adult bird; but there is still no sign of the magnificent orange side-plumes, which later still complete the attire of the perfect male. To effect these changes there must be at least three successive moultings; and as the birds were found by me in all the stages about the same time, it is probable that they moult only once a year, and that the full plumage is not acquired till the bird is four years old. It was long thought that the fine train of feathers was assumed for a short time only at the breeding season, but my own experience, as well as the observation of birds of an allied species which I brought home

with me, and which lived two years in this country, show
that the complete plumage is retained during the whole
year, except during a short period of moulting as with
most other birds.

The Great Bird of Paradise is very active and vigorous,
and seems to be in constant motion all day long. It is
very abundant, small flocks of females and young males
being constantly met with; and though the full-plumaged
birds are less plentiful, their loud cries, which are heard
daily, show that they also are very numerous. Their
note is, " Wawk-wawk-wawk—Wŏk, wŏk-wŏk," and is so
loud and shrill as to be heard a great distance, and to
form the most prominent and characteristic animal sound
in the Aru Islands. The mode of nidification is unknown; but the natives told me that the nest was formed
of leaves placed on an ant's nest, or on some projecting
limb of a very lofty tree, and they believe that it contains
only one young bird. The egg is quite unknown, and the
natives declared they had never seen it; and a very high
reward offered for one by a Dutch official did not meet
with success. They moult about January or February, and
in May, when they are in full plumage, the males assemble early in the morning to exhibit themselves in the
singular manner already described at p. 252. This habit
enables the natives to obtain specimens with comparative
ease. As soon as they find that the birds have fixed

upon a tree on which to assemble, they build a little shelter of palm leaves in a convenient place among the branches, and the hunter ensconces himself in it before daylight, armed with his bow and a number of arrows terminating in a round knob. A boy waits at the foot of the tree, and when the birds come at sunrise, and a sufficient number have assembled, and have begun to dance, the hunter shoots with his blunt arrow so strongly as to stun the bird, which drops down, and is secured and killed by the boy without its plumage being injured by a drop of blood. The rest take no notice, and fall one after another till some of them take the alarm. (See Frontispiece.)

The native mode of preserving them is to cut off the wings and feet, and then skin the body up to the beak, taking out the skull. A stout stick is then run up through the specimen coming out at the mouth. Round this some leaves are stuffed, and the whole is wrapped up in a palm spathe and dried in the smoky hut. By this plan the head, which is really large, is shrunk up almost to nothing, the body is much reduced and shortened, and the greatest prominence is given to the flowing plumage. Some of these native skins are very clean, and often have wings and feet left on ; others are dreadfully stained with smoke, and all give a most erroneous idea of the pro- portions of the living bird.

The Paradisea apoda, as far as we have any certain knowledge, is confined to the mainland of the Aru Islands, never being found in the smaller islands which surround the central mass. It is certainly not found in any of the parts of New Guinea visited by the Malay and Bugis traders, nor in any of the other islands where Birds of Paradise are obtained. But this is by no means conclusive evidence, for it is only in certain localities that the natives prepare skins, and in other places the same birds may be abundant without ever becoming known. It is therefore quite possible that this species may inhabit the great southern mass of New Guinea, from which Aru has been separated; while its near ally, which I shall next describe, is confined to the north-western peninsula.

The Lesser Bird of Paradise (Paradisea papuana of Bechstein), "Le petit Emeraude" of French authors, is a much smaller bird than the preceding, although very similar to it. It differs in its lighter brown colour, not becoming darker or purpled on the breast; in the extension of the yellow colour all over the upper part of the back and on the wing coverts; in the lighter yellow of the side plumes, which have only a tinge of orange, and at the tips are nearly pure white; and in the comparative shortness of the tail cirrhi. The female differs remarkably from the same sex in Paradisea apoda, by being entirely white on the under surface of the body, and is thus a much hand-somer bird. The young males are similarly coloured, and as they grow older they change to brown, and go through the same stages in acquiring the perfect plumage as has already been described in the allied species. It is this bird which is most commonly used in ladies' head-dresses in this country, and also forms an important article of commerce in the East.

The Paradisea papuana has a comparatively wide range, being the common species on the mainland of New Guinea, as well as on the islands of Mysol, Salwatty, Jobie, Biak and Sook. On the south coast of New Guinea, the Dutch naturalist, Muller, found it at the Oetanata river in longitude 136° E. I obtained it myself at Dorey; and the captain of the Dutch steamer *Etna* informed me that he had seen the feathers among the natives of Humboldt Bay, in 141° E. longitude. It is very probable, therefore, that it ranges over the whole of the mainland of New Guinea.

The true Paradise Birds are omnivorous, feeding on fruits and insects—of the former preferring the small figs; of the latter, grasshoppers, locusts, and phasmas, as well as cockroaches and caterpillars. When I returned home, in 1862, I was so fortunate as to find two adult males of this species in Singapore; and as they seemed healthy, and fed voraciously on rice, bananas, and cockroaches, I determined on giving the very high price asked for them—100*l.*—and to bring them to England by the overland route under my own care. On my way home I stayed a week at Bombay, to break the journey, and to lay in a fresh stock of bananas for my birds. I had great difficulty, however, in supplying them with insect food, for in the Peninsular and Oriental steamers cockroaches were scarce, and it was only by setting traps in the store-rooms, and by hunting an hour every night in the forecastle, that I could secure a few dozen of these creatures,—scarcely enough for a single meal. At Malta, where I stayed a fortnight, I got plenty of cockroaches from a bakehouse, and when I left, took with me several biscuit-tins' full, as provision for the voyage home. We came through the Mediterranean in March, with a very cold wind; and the only place on board the mail-steamer where their large cage could be accommodated was exposed to a strong current of air down a hatchway which stood open day and night, yet the birds never seemed to feel the cold. During the night journey from Marseilles to Paris it was a sharp frost; yet they arrived in London in perfect health, and lived in the Zoological Gardens for one, and two years, often displaying their beautiful plumes to the admiration of the spectators. It is evident, therefore, that the Paradise Birds are very hardy, and require air and exercise rather than heat; and I feel sure that if a good sized conservatory could be devoted to them, or if they could be turned loose

in the tropical department of the Crystal Palace or the
Great Palm House at Kew, they would live in this country
for many years.

The Red Bird of Paradise (Paradisea rubra of Viellot),
though allied to the two birds already described, is much
more distinct from them than they are from each other.
It is about the same size as Paradisea papuana (13 to 14
inches long), but differs from it in many particulars. The
side plumes, instead of being yellow, are rich crimson, and
only extend about three or four inches beyond the end of
the tail; they are somewhat rigid, and the ends are curved
downwards and inwards, and are tipped with white. The
two middle tail feathers, instead of being simply elongated
and deprived of their webs, are transformed into stiff black
ribands, a quarter of an inch wide, but curved like a split
quill, and resembling thin half cylinders of horn or whale-
bone. When a dead bird is laid on its back, it is seen that
these ribands take a curve or set, which brings them
round so as to meet in a double circle on the neck of the
bird; but when they hang downwards, during life, they
assume a spiral twist, and form an exceedingly graceful
double curve. They are about twenty-two inches long,
and always attract attention as the most conspicuous and
extraordinary feature of the species. The rich metallic
green colour of the throat extends over the front half of

the head to behind the eyes, and on the forehead forms a little double crest of scaly feathers, which adds much to the vivacity of the bird's aspect. The bill is gamboge yellow, and the iris blackish olive. (Figure at p. 353.)

The female of this species is of a tolerably uniform coffee-brown colour, but has a blackish head, and the nape, neck, and shoulders yellow, indicating the position of the brighter colours of the male. The changes of plumage follow the same order of succession as in the other species, the bright colours of the head and neck being first developed, then the lengthened filaments of the tail, and last of all, the red side plumes. I obtained a series of specimens, illustrating the manner in which the extraordinary black tail ribands are developed, which is very remarkable. They first appear as two ordinary feathers, rather shorter than the rest of the tail; the second stage would no doubt be that shown in a specimen of Paradisea apoda, in which the feathers are moderately lengthened, and with the web narrowed in the middle; the third stage is shown by a specimen which has part of the midrib bare, and terminated by a spatulate web; in another the bare midrib is a little dilated and semi-cylindrical, and the terminal web very small; in a fifth, the perfect black horny riband is formed, but it bears at its extremity a brown spatulate web, while in another a portion of the black riband itself bears, for a portion of its length, a narrow brown web. It is only after these changes are fully completed that the red side plumes begin to appear.

The successive stages of development of the colours and plumage of the Birds of Paradise are very interesting, from the striking manner in which they accord with the theory of their having been produced by the simple action of variation, and the cumulative power of selection by the females, of those male birds which were more than usually ornamental. Variations of *colour* are of all others the most frequent and the most striking, and are most easily modified and accumulated by man's selection of them. We should expect, therefore, that the sexual differences of *colour* would be those most early accumulated and fixed, and would therefore appear soonest in the young birds; and this is exactly what occurs in the Paradise Birds. Of all variations in the *form* of birds' feathers, none are so frequent as those in the head and tail. These occur more or less in every family of birds, and are easily produced in many domesticated varieties, while unusual developments of the feathers of the body are rare in the whole class of birds, and have seldom or never occurred in domesticated species. In accordance with these facts, we find the scale-formed plumes of the throat, the crests of the head, and the long cirrhi of the tail, all fully developed before the plumes which spring from the side of the body begin to make their appearance. If, on the other hand, the male

Paradise Birds have not acquired their distinctive plumage by successive variations, but have been as they are now from the moment they first appeared upon the earth, this succession becomes at the least unintelligible to us, for we can see no reason why the changes should not take place simultaneously, or in a reverse order to that in which they actually occur.

What is known of the habits of this bird, and the way in which it is captured by the natives, have already been described at page 362.

The Red Bird of Paradise offers a remarkable case of restricted range, being entirely confined to the small island of Waigiou, off the north-west extremity of New Guinea, where it replaces the allied species found in the other islands.

The three birds just described form a well-marked group, agreeing in every point of general structure, in their comparatively large size, the brown colour of their bodies, wings, and tail, and in the peculiar character of the ornamental plumage which distinguishes the male bird. The group ranges nearly over the whole area inhabited by the family of the Paradiseidæ, but each of the species has its own limited region, and is never found in the same district with either of its close allies. To these three birds properly belongs the generic title Paradisea, or true Paradise Bird.

The next species is the Paradisea regia of Linnæus, or
King Bird of Paradise, which differs so much from the
three preceding species as to deserve a distinct generic
name, and it has accordingly been called Cicinnurus regius.
By the Malays it is called " Burong rajah," or King Bird,
and by the natives of the Aru Islands " Goby-goby."

This lovely little bird is only about six and a half inches
long, partly owing to the very short tail, which does not
surpass the somewhat square wings.　The head, throat,
and entire upper surface are of the richest glossy crimson
red, shading to orange-crimson on the forehead, where the
feathers extend beyond the nostrils more than half-way
down the beak.　The plumage is excessively brilliant,
shining in certain lights with a metallic or glassy lustre.
The breast and belly are pure silky white, between which
colour and the red of the throat there is a broad band of
rich metallic green, and there is a small spot of the same
colour close above each eye.　From each side of the body
beneath the wing, springs a tuft of broad delicate feathers
about an inch and a half long, of an ashy colour, but
tipped with a broad band of emerald green, bordered
within by a narrow line of buff.　These plumes are concealed beneath the wing, but when the bird pleases, can be
raised and spread out so as to form an elegant semicircular
fan on each shoulder.　But another ornament still more
extraordinary, and if possible more beautiful, adorns this

little bird. The two middle tail feathers are modified into very slender wire-like shafts, nearly six inches long, each of which bears at the extremity, on the inner side only, a web of an emerald green colour, which is coiled up into a perfect spiral disc, and produces a most singular and charming effect. The bill is orange yellow, and the feet and legs of a fine cobalt blue. (See upper figure on the plate at the commencement of this chapter.)

The female of this little gem is such a plainly coloured bird, that it can at first sight hardly be believed to belong to the same species. The upper surface is of a dull earthy brown, a slight tinge of orange red appearing only on the margins of the quills. Beneath, it is of a paler yellowish brown, scaled and banded with narrow dusky markings. The young males are exactly like the female, and they no doubt undergo a series of changes as singular as those of Paradisea rubra; but, unfortunately, I was unable to obtain illustrative specimens.

This exquisite little creature frequents the smaller trees in the thickest parts of the forest, feeding on various fruits, often of a very large size for so small a bird. It is very active both on its wings and feet, and makes a whirring sound while flying, something like the South American manakins. It often flutters its wings and displays the beautiful fan which adorns its breast, while the star-bearing tail wires diverge in an elegant double curve. It

is tolerably plentiful in the Aru Islands, which led to its
being brought to Europe at an early period along with
Paradisea apoda. It also occurs in the island of Mysol,
and in every part of New Guinea which has been visited
by naturalists.

We now come to the remarkable little bird called the
" Magnificent," first figured by Buffon, and named Paradisea speciosa by Boddaert, which, with one allied species,
has been formed into a separate genus by Prince Buonaparte, under the name of Diphyllodes, from the curious
double mantle which clothes the back.

The head is covered with short brown velvety feathers,
which advance on the back so as to cover the nostrils
From the nape springs a dense mass of feathers of a straw-
yellow colour, and about one and a half inches long, forming a mantle over the upper part of the back. Beneath
this, and forming a band about one-third of an inch beyond it, is a second mantle of rich, glossy, reddish-brown
feathers. The rest of the back is orange-brown, the tail-
coverts and tail dark bronzy, the wings light orange-buff.
The whole under surface is covered with an abundance of
plumage springing from the margins of the breast, and
of a rich deep green colour, with changeable hues of
purple. Down the middle of the breast is a broad band
of scaly plumes of the same colour, while the chin and

D D 2

throat are of a rich metallic bronze. From the middle of the tail spring two narrow feathers of a rich steel blue, and about ten inches long. These are webbed on the inner side only, and curve outward, so as to form a double circle.

From what we know of the habits of allied species, we may be sure that the greatly developed plumage of this

THE MAGNIFICENT BIRD OF PARADISE. (*Diphyllodes speciosa.*)

bird is erected and displayed in some remarkable manner. The mass of feathers on the under surface are probably expanded into a hemisphere, while the beautiful yellow mantle is no doubt elevated so as to give the bird a very different appearance from that which it presents in the dried and flattened skins of the natives, through which alone it is at present known. The feet appear to be dark blue.

This rare and elegant little bird is found only on the mainland of New Guinea, and in the island of Mysol.

A still more rare and beautiful species than the last is the Diphyllodes wilsoni, described by Mr. Cassin from a native skin in the rich museum of Philadelphia. The same bird was afterwards named "Diphyllodes respublica" by Prince Buonaparte, and still later, "Schlegelia calva," by Dr. Bernstein, who was so fortunate as to obtain fresh specimens in Waigiou.

In this species the upper mantle is sulphur yellow, the lower one and the wings pure red, the breast plumes dark green, and the lengthened middle tail feathers much shorter than in the allied species. The most curious difference is, however, that the top of the head is bald, the bare skin being of a rich cobalt blue, crossed by several lines of black velvety feathers.

It is about the same size as Diphyllodes speciosa, and is no doubt entirely confined to the island of Waigiou. The female, as figured and described by Dr. Bernstein, is very like that of Cicinnurus regius, being similarly banded beneath; and we may therefore conclude that its near ally, the "Magnificent," is at least equally plain in this sex, of which specimens have not yet been obtained.

The Superb Bird of Paradise was first figured by Buffon, and was named by Boddaert, Paradisea atra, from

THE SUPERB BIRD OF PARADISE. (*Lophorina atra.*)

the black ground colour of its plumage. It forms the genus Lophorina of Viellot, and is one of the rarest and most brilliant of the whole group, being only known from mutilated native skins. This bird is a little larger than the Magnificent. The ground colour of the plumage is intense black, but with beautiful bronze reflections on the neck, and the whole head scaled with feathers of brilliant

metallic green and blue. Over its breast it bears a shield
formed of narrow and rather stiff feathers, much elongated
towards the sides, of a pure bluish-green colour, and with
a satiny gloss. But a still more extraordinary ornament
is that which springs from the back of the neck,—a shield
of a similar form to that on the breast, but much larger,
and of a velvety black colour, glossed with bronze and
purple. The outermost feathers of this shield are half
an inch longer than the wing, and when it is elevated
it must, in conjunction with the breast shield, completely
change the form and whole appearance of the bird. The
bill is black, and the feet appear to be yellow.

This wonderful little bird inhabits the interior of the
northern peninsula of New Guinea only. Neither I nor
Mr. Allen could hear anything of it in any of the islands
or on any part of the coast. It is true that it was obtained
from the coast-natives by Lesson; but when at Sorong in
1861, Mr. Allen learnt that it is only found three days'
journey in the interior. Owing to these " Black Birds of
Paradise," as they are called, not being so much valued as
articles of merchandise, they now seem to be rarely preserved by the natives, and it thus happened that during
several years spent on the coasts of New Guinea and in
the Moluccas I was never able to obtain a skin. We are
therefore quite ignorant of the habits of this bird, and
also of its female, though the latter is no doubt as plain

and inconspicuous as in all the other species of this
family.

The Golden, or Six-shafted Paradise Bird, is another
rare species, first figured by Buffon, and never yet

THE SIX-SHAFTED BIRD OF PARADISE. (*Parotia sexpennis.*)

obtained in perfect condition. It was named by Boddaert, Paradisea sexpennis, and forms the genus Parotia
of Viellot. This wonderful bird is about the size
of the female Paradisea rubra. The plumage appears
at first sight black, but it glows in certain lights
with bronze and deep purple. The throat and breast
are scaled with broad flat feathers of an intense golden

hue, changing to green and blue tints in certain lights. On the back of the head is a broad recurved band of feathers, whose brilliancy is indescribable, resembling the sheen of emerald and topaz rather than any organic substance. Over the forehead is a large patch of pure white feathers, which shine like satin; and from the sides of the head spring the six wonderful feathers from which the bird receives its name. These are slender wires, six inches long, with a small oval web at the extremity. In addition to these ornaments, there is also an immense tuft of soft feathers on each side of the breast, which when elevated must entirely hide the wings, and give the bird an appearance of being double its real bulk. The bill is black, short, and rather compressed, with the feathers advancing over the nostrils, as in Cicinnurus regius. This singular and brilliant bird inhabits the same region as the Superb Bird of Paradise, and nothing whatever is known about it but what we can derive from an examination of the skins preserved by the natives of New Guinea.

The Standard Wing, named Semioptera wallacei by Mr. G. R. Gray, is an entirely new form of Bird of Paradise, discovered by myself in the island of Batchian, and especially distinguished by a pair of long narrow feathers of a white colour, which spring from among the short plumes which clothe the bend of the wing, and are capable of being erected at pleasure. The general colour of this bird is a delicate olive-brown, deepening to a kind of bronzy olive in the middle of the back, and changing to a delicate ashy violet with a metallic gloss, on the crown of the head. The feathers, which cover the nostrils and extend half-way down the beak, are loose and curved upwards. Beneath, it is much more beautiful. The scale-like feathers of the breast are margined with rich metallic blue-green, which colour entirely covers the throat and sides of the neck, as well as the long pointed plumes which spring from the sides of the breast, and extend nearly as far as the end of the wings. The most curious feature of the bird, however, and one altogether unique in the whole class, is found in the pair of long narrow delicate feathers which spring from each wing close to the bend. On lifting the wing-coverts they are seen to arise from two tubular horny sheaths, which diverge from near the point of junction of the carpal bones. As already described at p. 41, they are erectile, and when the bird is excited are spread out at right angles to the wing and slightly divergent. They are from six to six and a half inches long, the upper one slightly exceeding the lower. The total length of the bird is eleven inches. The bill is horny olive, the iris deep olive, and the feet bright orange.

The female bird is remarkably plain, being entirely of a dull pale earthy brown, with only a slight tinge of ashy

violet on the head to relieve its general monotony; and
the young males exactly resemble her.   (See figures at
p. 41.)

This bird frequents the lower trees of the forests, and,
like most Paradise Birds, is in constant motion—flying
from branch to branch, clinging to the twigs and even to
the smooth and vertical trunks almost as easily as a wood-
pecker.   It continually utters a harsh, creaking note,
somewhat intermediate between that of Paradisea apoda
and the more musical cry of Cicinnurus regius.   The
males at short intervals open and flutter their wings, erect
the long shoulder feathers, and spread out the elegant
green breast shields.

The Standard Wing is found in Gilolo as well as in
Batchian, and all the specimens from the former island
have the green breast shield rather longer, the crown of the
head darker violet, and the lower parts of the body rather
more strongly scaled with green.   This is the only Paradise
Bird yet found in the Moluccan district, all the others
being confined to the Papuan Islands and North
Australia.

We now come to the Epimachidæ, or Long-billed Birds
of Paradise, which, as before stated, ought not to be separated from the Paradiseidæ by the intervention of any
other birds.   One of the most remarkable of these is the

Twelve-wired Paradise Bird, Paradisea alba of Blumenbach, but now placed in the genus Seleucides of Lesson.

This bird is about twelve inches long, of which the compressed and curved beak occupies two inches. The colour of the breast and upper surface appears at first sight nearly black, but a close examination shows that no part of it is devoid of colour; and by holding it in various lights, the most rich and glowing tints become visible. The head, covered with short velvety feathers, which advance on the chin much further than on the upper part of the beak, is of a purplish bronze colour; the whole of the back and shoulders is rich bronzy green, while the closed wings and tail are of the most brilliant violet purple, all the plumage having a delicate silky gloss. The mass of feathers which cover the breast is really almost black, with faint glosses of green and purple, but their outer edges are margined with glittering bands of emerald green. The whole lower part of the body is rich buffy yellow, including the tuft of plumes which spring from the sides, and extend an inch and a half beyond the tail. When skins are exposed to the light the yellow fades into dull white, from which circumstance it derived its specific name. About six of the innermost of these plumes on each side have the midrib elongated into slender black wires, which bend at right angles, and curve somewhat backwards to a length of about ten inches, forming one of those extraordinary and fantastic ornaments with which this group of birds abounds. The bill is jet black, and the feet bright yellow. (See lower figure on the plate at the beginning of this chapter).

The female, although not quite so plain a bird as in some other species, presents none of the gay colours or ornamental plumage of the male. The top of the head and back of the neck are black, the rest of the upper parts rich reddish brown; while the under surface is entirely yellowish ashy, somewhat blackish on the breast, and crossed throughout with narrow blackish wavy bands.

The Seleucides alba is found in the island of Salwatty, and in the north-western parts of New Guinea, where it frequents flowering trees, especially sago-palms and pandani, sucking the flowers, round and beneath which its unusually large and powerful feet enable it to cling. Its motions are very rapid. It seldom rests more than a few moments on one tree, after which it flies straight off, and with great swiftness, to another. It has a loud shrill cry, to be heard a long way, consisting of " Cáh, cáh," repeated five or six times in a descending scale, and at the last note it generally flies away. The males are quite solitary in their habits, although, perhaps, they assemble at certain times like the true Paradise Birds. All the specimens shot and opened by my assistant Mr. Allen, who obtained this fine bird during his last voyage to New Guinea, had nothing in their stomachs but a brown sweet liquid, probably the nectar of the flowers on which they had been feeding. They certainly, however, eat both fruit and insects, for a specimen which I saw alive on board a Dutch steamer ate cockroaches and papaya fruit voraciously. This bird had the curious habit of resting at noon with the bill pointing vertically upwards. It died on the passage to Batavia, and I secured the body and formed a skeleton, which shows indisputably that it is really a Bird of Paradise. The tongue is very long and extensible, but flat and a little fibrous at the end, exactly like the true Paradiseas.

In the island of Salwatty, the natives search in the forests till they find the sleeping place of this bird, which they know by seeing its dung upon the ground. It is generally in a low bushy tree. At night they climb up the tree, and either shoot the birds with blunt arrows, or even catch them alive with a cloth. In New Guinea they are caught by placing snares on the trees frequented by them, in the same way as the Red Paradise Birds are caught in Waigiou, and which has already been described at page 362.

The great Epimaque, or Long-tailed Paradise Bird (Epimachus magnus), is another of these wonderful creatures, only known by the imperfect skins prepared by the natives. In its dark velvety plumage, glossed with bronze and purple, it resembles the Seleucides alba, but it bears a magnificent tail more than two feet long, glossed on the upper surface with the most intense opalescent blue. Its chief ornament, however, consists in the group of broad plumes which spring from the sides of the breast, and which are dilated at the extremity, and banded with the most vivid metallic blue and green. The bill is long and curved, and the feet black, and

THE LONG-TAILED BIRD OF PARADISE.
(*Epimachus magnus.*)

similar to those of the allied forms. The total length of this fine bird is between three and four feet.

This splendid bird inhabits the mountains of New Guinea, in the same district with the Superb and the Six-shafted Paradise Birds, and I was informed is sometimes found in the ranges near the coast. I was several times assured by different natives that this bird makes its nest in a hole under ground, or under rocks, always choosing a place with two apertures, so that it may enter at one and go out at the other. This is very unlike what we should suppose to be the habits of the bird, but it is not easy to conceive how the story originated if it is not true; and all travellers know that native accounts of the habits of animals, however strange they may seem, almost invariably turn out to be correct.

The Scale-breasted Paradise Bird (Epimachus magnificus of Cuvier) is now generally placed with the Australian Rifle birds in the genus Ptiloris. Though very beautiful, these birds are less strikingly decorated with accessory plumage than the other species we have been describing, their chief ornament being a more or less developed breastplate of stiff metallic green feathers, and a small tuft of somewhat hairy plumes on the sides of the breast. The back and wings of this species are of an intense velvety black, faintly glossed in certain lights with rich purple. The two broad middle tail feathers are opalescent green-blue with a velvety surface, and the top of the head is covered with feathers resembling scales of burnished steel. A large triangular space covering the chin, throat, and breast, is densely scaled with feathers, having a steel-blue or green lustre, and a silky feel. This is edged below with a narrow band of black, followed by shiny bronzy green, below which the body is covered with hairy feathers of a

rich claret colour, deepening to black at the tail.   The tufts
of side plumes somewhat resemble those of the true Birds
of Paradise, but are scanty, about as long as the tail, and of
a black colour.   The sides of the head are rich violet, and
velvety feathers extend on each side of the beak over the
nostrils.

I obtained at Dorey a young male of this bird, in a state
of plumage which is no doubt that of the adult female, as
is the case in all the allied species.   The upper surface,
wings, and tail are rich reddish brown, while the under
surface is of a pale ashy colour, closely barred throughout
with narrow wavy black bands.   There is also a pale
banded stripe over the eye, and a long dusky stripe from
the gape down each side of the neck.   This bird is fourteen inches long, whereas the native skins of the adult
male are only about ten inches, owing to the way in which
the tail is pushed in, so as to give as much prominence as
possible to the ornamental plumage of the breast.

At Cape York, in North Australia, there is a closely
allied species, Ptiloris alberti, the female of which is very
similar to the young male bird here described.   The beautiful Rifle Birds of Australia, which much resemble these
Paradise Birds, are named Ptiloris paradiseus and Ptiloris
victoriæ.   The Scale-breasted Paradise Bird seems to be
confined to the mainland of New Guinea, and is less rare
than several of the other species.

There are three other New Guinea birds which are by some authors classed with the Birds of Paradise, and which, being almost equally remarkable for splendid plumage, deserve to be noticed here. The first is the Paradise pie (Astrapia nigra of Lesson), a bird of the size of Paradisea rubra, but with a very long tail, glossed above with intense violet. The back is bronzy black, the lower parts green, the throat and neck bordered with loose broad feathers of an intense coppery hue, while on the top of the head and neck they are glittering emerald green. All the plumage round the head is lengthened and erectile, and when spread out by the living bird must have an effect hardly surpassed by any of the true Paradise Birds. The bill is black and the feet yellow. The Astrapia seems to me to be somewhat intermediate between the Paradiseidæ and Epimachidæ.

There is an allied species, having a bare carunculated head, which has been called Paradigalla carunculata. It is believed to inhabit, with the preceding, the mountainous interior of New Guinea, but is exceedingly rare, the only known specimen being in the Philadelphia Museum.

The Paradise Oriole is another beautiful bird, which is now sometimes classed with the Birds of Paradise. It has been named Paradisea aurea and Oriolus aureus by the old naturalists, and is now generally placed in the same genus

as the Regent Bird of Australia (Sericulus chrysocephalus).
But the form of the bill and the character of the plumage
seem to me to be so different that it will have to form
a distinct genus. This bird is almost entirely yellow,
with the exception of the throat, the tail, and part of the
wings and back, which are black; but it is chiefly characterised by a quantity of long feathers of an intense glossy
orange colour, which cover its neck down to the middle of
the back, almost like the hackles of a game-cock.

This beautiful bird inhabits the mainland of New
Guinea, and is also found in Salwatty, but is so rare that I
was only able to obtain one imperfect native skin, and
nothing whatever is known of its habits.

I will now give a list of all the Birds of Paradise yet
known, with the places they are believed to inhabit.

1. Paradisea apoda (The Great Paradise Bird). Aru Islands.

2. Paradisea papuana (The Lesser Paradise Bird). New Guinea, Mysol,
Jobie.

3. Paradisea rubra (The Red Paradise Bird). Waigiou.

4. Cicinnurus regius (The King Paradise Bird). New Guinea, Aru
Islands, Mysol, Salwatty.

5. Diphyllodes speciosa (The Magnificent). New Guinea, Mysol, Salwatty.

6. Diphyllodes wilsoni (The Red Magnificent). Waigiou.

7. Lophorina atra (The Superb). New Guinea.

8. Parotia sexpennis (The Golden Paradise Bird). New Guinea.

9. Semioptera wallacei (The Standard Wing). Batchian, Gilolo.

10. Epimachus magnus (The Long-tailed Paradise Bird). New Guinea.

11. Seleucides alba (The Twelve-wired Paradise Bird). New Guinea,
Salwatty.

E E 2

12. Ptiloris magnifica (The Scale-breasted Paradise Bird). New Guinea.

13. Ptiloris alberti (Prince Albert's Paradise Bird). North Australia.

14. Ptiloris paradisea (The Rifle Bird). East Australia.

15. Ptiloris victoriæ (The Victorian Rifle Bird). North-East Australia.

16. Astrapia nigra (The Paradise Pie). New Guinea.

17. Paradigalla carunculata (The Carunculated Paradise Pie). New Guinea.

18. (?) Sericulus aureus (The Paradise Oriole). New Guinea, Salwatty.

We see, therefore, that of the eighteen species which seem to deserve a place among the Birds of Paradise, eleven are known to inhabit the great island of New Guinea, eight of which are entirely confined to it and the hardly separated island of Salwatty. But if we consider those islands which are now united to New Guinea by a shallow sea to really form a part of it, we shall find that fourteen of the Paradise Birds belong to that country, while three inhabit the northern and eastern parts of Australia, and one the Moluccas. All the more extraordinary and magnificent species are, however, entirely confined to the Papuan region.

Although I devoted so much time to a search after these wonderful birds, I only succeeded myself in obtaining five species during a residence of many months in the Aru Islands, New Guinea, and Waigiou. Mr. Allen's voyage to Mysol did not procure a single additional species, but we both heard of a place called Sorong, on the mainland of New Guinea, near Salwatty, where we were told that all the kinds we desired could be obtained. We therefore determined that he should visit this place, and endeavour to penetrate into the interior among the natives, who actually shoot and skin the Birds of Paradise. He went in the small prau I had fitted up at Goram, and through the kind assistance of the Dutch Resident at Ternate, a lieutenant and two soldiers were sent by the Sultan of Tidore to accompany and protect him, and to assist him in getting men and in visiting the interior.

Notwithstanding these precautions, Mr. Allen met with difficulties in this voyage which we had neither of us encountered before. To understand these, it is necessary to consider that the Birds of Paradise are an article of commerce, and are the monopoly of the chiefs of the coast villages, who obtain them at a low rate from the mountaineers, and sell them to the Bugis traders. A portion is also paid every year as tribute to the Sultan of Tidore. The natives are therefore very jealous of a stranger, especially a European, interfering in their trade, and above all of going into the interior to deal with the mountaineers themselves. They of course think he will raise the prices in the interior, and lessen the supply on the coast, greatly to their disadvantage; they also think their tribute will be raised if a European takes back a quantity of the rare sorts; and they have besides a vague

and very natural dread of some ulterior object in a
white man's coming at so much trouble and expense to
their country only to get Birds of Paradise, of which
they know he can buy plenty (of the common yellow
ones which alone they value) at Ternate, Macassar, or
Singapore.

It thus happened that when Mr. Allen arrived at Sorong,
and explained his intention of going to seek Birds of
Paradise in the interior, innumerable objections were
raised. He was told it was three or four days' journey
over swamps and mountains; that the mountaineers were
savages and cannibals, who would certainly kill him;
and, lastly, that not a man in the village could be found
who dare go with him. After some days spent in these
discussions, as he still persisted in making the attempt,
and showed them his authority from the Sultan of Tidore
to go where he pleased and receive every assistance, they
at length provided him with a boat to go the first part
of the journey up a river; at the same time, however,
they sent private orders to the interior villages to refuse
to sell any provisions, so as to compel him to return. On
arriving at the village where they were to leave the river
and strike inland, the coast people returned, leaving Mr.
Allen to get on as he could. Here he called on the
Tidore lieutenant to assist him, and procure men as
guides and to carry his baggage to the villages of the

mountaineers. This, however, was not so easily done. A quarrel took place, and the natives, refusing to obey the imperious orders of the lieutenant, got out their knives and spears to attack him and his soldiers ; and Mr. Allen himself was obliged to interfere to protect those who had come to guard him. The respect due to a white man and the timely distribution of a few presents prevailed ; and, on showing the knives, hatchets, and beads he was willing to give to those who accompanied him, peace was restored, and the next day, travelling over a frightfully rugged country, they reached the villages of the mountaineers. Here Mr. Allen remained a month without any interpreter through whom he could understand a word or communicate a want. However, by signs and presents and a pretty liberal barter, he got on very well, some of them accompanying him every day in the forest to shoot, and receiving a small present when he was successful.

In the grand matter of the Paradise Birds, however, little was done. Only one additional species was found, the Seleucides alba, of which he had already obtained a specimen in Salwatty; but he learnt that the other kinds, of which he showed them drawings, were found two or three days' journey farther in the interior. When I sent my men from Dorey to Amberbaki, they heard exactly the same story—that the rarer sorts were only found several days' journey in the interior, among rugged mountains, and that the skins were prepared by savage tribes who had never even been seen by any of the coast people.

It seems as if Nature had taken precautions that these her choicest treasures should not be made too common, and thus be undervalued. This northern coast of New Guinea is exposed to the full swell of the Pacific Ocean, and is rugged and harbourless. The country is all rocky and mountainous, covered everywhere with dense forests, offering in its swamps and precipices and serrated ridges an almost impassable barrier to the unknown interior; and the people are dangerous savages, in the very lowest stage of barbarism. In such a country, and among such a people, are found these wonderful productions of Nature, the Birds of Paradise, whose exquisite beauty of form and colour and strange developments of plumage are calculated to excite the wonder and admiration of the most civilized and the most intellectual of mankind, and to furnish inexhaustible materials for study to the naturalist, and for speculation to the philosopher.

Thus ended my search after these beautiful birds. Five voyages to different parts of the district they inhabit, each occupying in its preparation and execution the larger part of a year, produced me only five species out of the fourteen

known to exist in the New Guinea district. The kinds
obtained are those that inhabit the coasts of New Guinea
and its islands, the remainder seeming to be strictly confined to the central mountain-ranges of the northern
peninsula ; and our researches at Dorey and Amberbaki,
near one end of this peninsula, and at Salwatty and
Sorong, near the other, enable me to decide with some
certainty on the native country of these rare and lovely
birds, good specimens of which have never yet been seen
in Europe.

It must be considered as somewhat extraordinary that,
during five years' residence and travel in Celebes, the
Moluccas, and New Guinea, I should never have been
able to purchase skins of half the species which Lesson,
forty years ago, obtained during a few weeks in the
same countries. I believe that all, except the common
species of commerce, are now much more difficult to obtain
than they were even twenty years ago ; and I impute it
principally to their having been sought after by the Dutch
officials through the Sultan of Tidore. The chiefs of the
annual expeditions to collect tribute have had orders to get
all the rare sorts of Paradise Birds ; and as they pay little
or nothing for them (it being sufficient to say they are for
the Sultan), the head men of the coast villages would
for the future refuse to purchase them from the mountaineers, and confine themselves instead to the commoner

species, which are less sought after by amateurs, but are a
more profitable merchandise.   The same causes frequently
lead the inhabitants of uncivilized countries to conceal
minerals or other natural products with which they may
become acquainted, from the fear of being obliged to pay
increased tribute, or of bringing upon themselves a new
and oppressive labour.

CHAPTER XXXIX.

NEW GUINEA, with the islands joined to it by a shallow sea, constitute the Papuan group, characterised by a very close resemblance in their peculiar forms of life. Having already, in my chapters on the Aru Islands and on the Birds of Paradise, given some details of the natural history of this district, I shall here confine myself to a general sketch of its animal productions, and of their relations to those of the rest of the world.

New Guinea is perhaps the largest island on the globe, being a little larger than Borneo. It is nearly fourteen hundred miles long, and in the widest part four hundred broad, and seems to be everywhere covered with luxuriant forests. Almost everything that is yet known of its natural productions comes from the north-western peninsula, and a few islands grouped around it. These do not constitute a tenth part of the area of the whole island, and are so cut off from it, that their fauna may well be somewhat different; yet they have produced us (with a very partial exploration) no less than two hundred and fifty species of land birds, almost all unknown elsewhere, and comprising some of the most curious and most beautiful of the feathered tribes.   It is needless to say how much interest attaches to the far larger unknown portion of this great island, the greatest *terra incognita* that still remains for the naturalist to explore, and the only region where altogether new and unimagined forms of life may perhaps be found.   There is now, I am happy to say, some chance that this great country will no longer remain absolutely unknown to us.   The Dutch Government have granted a well-equipped steamer to carry a naturalist (Mr. Rosenberg, already mentioned in this work) and assistants to New Guinea, where they are to spend some years in circumnavigating the island, ascending its large rivers as far as possible into the interior, and making extensive collections of its natural productions.

The Mammalia of New Guinea and the adjacent islands, yet discovered, are only seventeen in number.   Two of these are bats, one is a pig of a peculiar species (Sus papuensis), and the rest are all marsupials.   The bats are, no doubt, much more numerous, but there is every reason to believe that whatever new land Mammalia may be discovered will belong to the marsupial order.   One of these is a true kangaroo, very similar to some of the middle-sized kangaroos of Australia, and it is remarkable as being the first animal of the kind ever seen by Euro-

peans. It inhabits Mysol and the Aru Islands (an allied
species being found in New Guinea), and was seen and
described by Le Brun in 1714, from living specimens at
Batavia. A much more extraordinary creature is the tree-
kangaroo, two species of which are known from New
Guinea. These animals do not differ very strikingly in form
from the terrestrial kangaroos, and appear to be but imperfectly adapted to an arboreal life, as they move rather
slowly, and do not seem to have a very secure footing on
the limb of a tree. The leaping power of the muscular tail
is lost, and powerful claws have been acquired to assist
in climbing, but in other respects the animal seems better
adapted to walk on *terra firma*. This imperfect adaptation may be due to the fact of there being no carnivora
in New Guinea, and no enemies of any kind from which
these animals have to escape by rapid climbing. Four
species of Cuscus, and the small flying opossum, also inhabit New Guinea ; and there are five other smaller marsupials, one of which is the size of a rat, and takes its
place by entering houses and devouring provisions.

The birds of New Guinea offer the greatest possible
contrast to the Mammalia, since they are more numerous,
more beautiful, and afford more new, curious, and elegant
forms than those of any other island on the globe.
Besides the Birds of Paradise, which we have already

sufficiently considered, it possesses a number of other
curious birds, which in the eyes of the ornithologist
almost serves to distinguish it as one of the primary
divisions of the earth. Among its thirty species of
parrots are the Great Black Cockatoo, and the little rigid-
tailed Nasiterna, the giant and the dwarf of the whole
tribe. The bare-headed Dasyptilus is one of the most
singular parrots known; while the beautiful little long-
tailed Charmosyna, and the great variety of gorgeously-
coloured lories, have no parallels elsewhere. Of pigeons
it possesses about forty distinct species, among which are
the magnificent crowned pigeons, now so well known in
our aviaries, and pre-eminent both for size and beauty;
the curious Trugon terrestris, which approaches the still
more strange Didunculus of Samoa ; and a new genus
(Henicophaps), discovered by myself, which possesses a
very long and powerful bill, quite unlike that of any other
pigeon. Among its sixteen kingfishers, it possesses the
curious hook-billed Macrorhina, and a red and blue
Tanysiptera, the most beautiful of that beautiful genus.
Among its perching birds are the fine genus of crow-like
starlings, with brilliant plumage (Manucodia) ; the curious
pale-coloured crow (Gymnocorvus senex) ; the abnormal
red and black flycatcher (Peltops blainvillii) ; the curious
little boat-billed flycatchers (Machærirhynchus) ; and the
elegant blue flycatcher-wrens (Todopsis).

The naturalist will obtain a clearer idea of the variety
and interest of the productions of this country, by the
statement, that its land birds belong to 108 genera, of
which 29 are exclusively characteristic of it; while 35
belong to that limited area which includes the Moluccas
and North Australia, and whose species of these genera
have been entirely derived from New Guinea. About
one-half of the New Guinea genera are found also in
Australia, about one-third in India and the Indo-Malay
islands.

A very curious fact, not hitherto sufficiently noticed, is
the appearance of a pure Malay element in the birds of New
Guinea. We find two species of Eupetes, a curious Malayan
genus allied to the forked-tail water-chats; two of Alcippe,
an Indian and Malay wren-like form; an Arachnothera,
quite resembling the spider-catching honeysuckers of Malacca; two species of Gracula, the Mynahs of India; and
a curious little black Prionochilus, a saw-billed fruit-
pecker, undoubtedly allied to the Malayan form, although
perhaps a distinct genus. Now not one of these birds, or
anything allied to them, occurs in the Moluccas, or (with
one exception) in Celebes or Australia; and as they are
most of them birds of short flight, it is very difficult to
conceive how or when they could have crossed the space
of more than a thousand miles, which now separates them
from their nearest allies. Such facts point to changes

of land and sea on a large scale, and at a rate which, measured by the time required for a change of species, must be termed rapid. By speculating on such changes, we may easily see how partial waves of immigration may have entered New Guinea, and how all trace of their passage may have been obliterated by the subsequent disappearance of the intervening land.

There is nothing that the study of geology teaches us that is more certain or more impressive than the extreme instability of the earth's surface. Everywhere beneath our feet we find proofs that what is land has been sea, and that where oceans now spread out has once been land; and that this change from sea to land, and from land to sea, has taken place, not once or twice only, but again and again, during countless ages of past time. Now the study of the distribution of animal life upon the present surface of the earth, causes us to look upon this constant interchange of land and sea—this making and unmaking of continents, this elevation and disappearance of islands—as a potent reality, which has always and everywhere been in progress, and has been the main agent in determining the manner in which living things are now grouped and scattered over the earth's surface. And when we continually come upon such little anomalies of distribution as that just now described, we find the only rational explanation of them, in those repeated elevations and depressions which have left their record in mysterious, but still intelligible characters on the face of organic nature.

The insects of New Guinea are less known than the birds, but they seem almost equally remarkable for fine forms and brilliant colours. The magnificent green and yellow Ornithopteræ are abundant, and have most probably spread westward from this point as far as India. Among the smaller butterflies are several peculiar genera of Nymphalidæ and Lycænidæ, remarkable for their large size, singular markings, or brilliant coloration. The largest and most beautiful of the clear-winged moths (Cocytia d'urvillei) is found here, as well as the large and handsome green moth (Nyctalemon orontes). The beetles furnish us with many species of large size, and of the most brilliant metallic lustre, among which the Tmesisternus mirabilis, a longicorn beetle of a golden green colour; the excessively brilliant rose-chafers, Lomaptera wallacei and Anacamptorhina fulgida; one of the handsomest of the Buprestidæ, Calodema wallacei; and several fine blue weevils of the genus Eupholus, are perhaps the most conspicuous. Almost all the other orders furnish us with large or extraordinary forms. The curious horned flies have already been mentioned; and among the Orthoptera the great shielded grasshoppers are the most remarkable. The species here figured (Mega-

lodon ensifer) has the thorax covered by a large triangular
horny shield, two and a half inches long, with serrated
edges, a somewhat wavy, hollow surface, and a faint
median line, so as very closely to resemble a leaf. The
glossy wing-coverts (when fully expanded, more than nine

THE GREAT-SHIELDED GRASSHOPPER.

inches across) are of a fine green colour and so beautifully
veined as to imitate closely some of the large shining
tropical leaves. The body is short, and terminated in the
female by a long curved sword-like ovipositor (not seen
in the cut), and the legs are all long and strongly-spined.
These insects are sluggish in their motions, depending

for safety on their resemblance to foliage, their horny shield and wing-coverts, and their spiny legs.

The large islands to the east of New Guinea are very little known, but the occurrence of crimson lories, which are quite absent from Australia, and of cockatoos allied to those of New Guinea and the Moluccas, shows that they belong to the Papuan group ; and we are thus able to define the Malay Archipelago as extending eastward to the Solomon's Islands. New Caledonia and the New Hebrides, on the other hand, seem more nearly allied to Australia ; and the rest of the islands of the Pacific, though very poor in all forms of life, possess a few peculiarities which compel us to class them as a separate group. Although as a matter of convenience I have always separated the Moluccas as a distinct zoological group from New Guinea, I have at the same time pointed out that its fauna was chiefly derived from that island, just as that of Timor was chiefly derived from Australia. If we were dividing the Australian region for zoological purposes alone, we should form three great groups : one comprising Australia, Timor, and Tasmania ; another New Guinea, with the islands from Bouru to the Solomon's group ; and the third comprising the greater part of the Pacific Islands.

The relation of the New Guinea fauna to that of Australia is very close. It is best marked in the Mammalia by the abundance of marsupials, and the almost complete absence of all other terrestrial forms. In birds it is less striking, although still very clear, for all the remarkable old-world forms which are absent from the one are equally so from the other, such as Pheasants, Grouse, Vultures, and Woodpeckers; while Cockatoos, Broad-tailed Parrots, Podargi, and the great families of the Honey-suckers and Brush-turkeys, with many others, comprising no less than twenty-four genera of land-birds, are common to both countries, and are entirely confined to them.

When we consider the wonderful dissimilarity of the two regions in all those physical conditions which were once supposed to determine the forms of life—Australia, with its open plains, stony deserts, dried up rivers, and changeable temperate climate; New Guinea, with its luxuriant forests, uniformly hot, moist, and evergreen—this great similarity in their productions is almost astounding, and unmistakeably points to a common origin. The resemblance is not nearly so strongly marked in insects, the reason obviously being, that this class of animals are much more immediately dependent on vegetation and climate than are the more highly organized birds and Mammalia. Insects also have far more effective means of distribution, and have spread widely into every district favourable to their development and increase. The giant Ornithopteræ have thus spread from New Guinea over the whole Archipelago, and as far as the base of the Himalayas; while the

elegant long-horned Anthribidæ have spread in the opposite
direction from Malacca to New Guinea, but owing to unfavourable conditions have not been able to establish
themselves in Australia. That country, on the other hand,
has developed a variety of flower-haunting Chafers and
Buprestidæ, and numbers of large and curious terrestrial
Weevils, scarcely any of which are adapted to the damp
gloomy forests of New Guinea, where entirely different
forms are to be found. There are, however, some groups of
insects, constituting what appear to be the remains of the
ancient population of the equatorial parts of the Australian
region, which are still almost entirely confined to it. Such
are the interesting sub-family of Longicorn coleoptera—
Tmesisternitæ; one of the best-marked genera of Buprestidæ—Cyphogastra; and the beautiful weevils forming the
genus Eupholus. Among butterflies we have the genera
Mynes, Hypocista, and Elodina, and the curious eye-
spotted Drusilla, of which last a single species is found in
Java, but in no other of the western islands.

The facilities for the distribution of plants are still
greater than they are for insects, and it is the opinion of
eminent botanists, that no such clearly-defined regions can
be marked out in botany as in zoology. The causes which
tend to diffusion are here most powerful, and have led to
such intermingling of the floras of adjacent regions that
none but broad and general divisions can now be detected.

These remarks have an important bearing on the problem
of dividing the surface of the earth into great regions, distinguished by the radical difference of their natural productions.   Such difference we now know to be the direct
result of long-continued separation by more or less impassable barriers ; and as wide oceans and great contrasts
of temperature are the most complete barriers to the
dispersal of all terrestrial forms of life, the primary
divisions of the earth should in the main serve for all
terrestrial organisms.   However various may be the effects
of climate, however unequal the means of distribution,
these will never altogether obliterate the radical effects of
long-continued isolation; and it is my firm conviction, that
when the botany and the entomology of New Guinea and
the surrounding islands become as well known as are
their mammals and birds, these departments of nature
will also plainly indicate the radical distinctions of the
Indo-Malayan and Austro-Malayan regions of the great
Malay Archipelago.

CHAPTER XI.

I PROPOSE to conclude this account of my Eastern travels, with a short statement of my views as to the races of man which inhabit the various parts of the Archipelago, their chief physical and mental characteristics, their affinities with each other and with surrounding tribes, their migrations, and their probable origin.

Two very strongly contrasted races inhabit the Archipelago—the Malays, occupying almost exclusively the larger western half of it, and the Papuans, whose head-quarters are New Guinea and several of the adjacent islands. Between these in locality, are found tribes who are also intermediate in their chief characteristics, and it is sometimes a nice point to determine whether they belong to one or the other race, or have been formed by a mixture of the two.

The Malay is undoubtedly the most important of these two races, as it is the one which is the most civilized, which has come most into contact with Europeans, and

which alone has any place in history   What may be called
the true Malay races, as distinguished from others who
have merely a Malay element in their language, present a
considerable uniformity of physical and mental characteristics, while there are very great differences of civilization and of language.   They consist of four great, and a few
minor semi-civilized tribes, and a number of others who
may be termed savages.   The Malays proper inhabit the
Malay peninsula, and almost all the coast regions of
Borneo and Sumatra.   They all speak the Malay language,
or dialects of it; they write in the Arabic character, and
are Mahometans in religion.   The Javanese inhabit Java,
part of Sumatra, Madura, Bali, and part of Lombock.
They speak the Javanese and Kawi languages, which
they write in a native character.   They are now Mahometans in Java, but Brahmins in Bali and Lombock.   The
Bugis are the inhabitants of the greater parts of Celebes,
and there seems to be an allied people in Sumbawa.   They
speak the Bugis and Macassar languages, with dialects, and
have two different native characters in which they write
these.   They are all Mahometans.   The fourth great race
is that of the Tagalas in the Philippine Islands, about
whom, as I did not visit those Islands, I shall say
little.   Many of them are now Christians, and speak
Spanish as well as their native tongue, the Tagala.   The
Moluccan-Malays, who inhabit chiefly Ternate, Tidore,

Batchian, and Amboyna, may be held to form a fifth
division of semi-civilized Malays. They are all Mahometans, but they speak a variety of curious languages,
which seem compounded of Bugis and Javanese, with the
languages of the savage tribes of the Moluccas.

The savage Malays are the Dyaks of Borneo; the
Battaks and other wild tribes of Sumatra; the Jakuns of
the Malay Peninsula; the aborigines of Northern Celebes,
of the Sula island, and of part of Bouru.

The colour of all these varied tribes is a light reddish
brown, with more or less of an olive tinge, not varying in
any important degree over an extent of country as large as
all Southern Europe. The hair is equally constant, being
invariably black and straight, and of a rather coarse texture, so that any lighter tint, or any wave or curl in it,
is an almost certain proof of the admixture of some foreign
blood. The face is nearly destitute of beard, and the
breast and limbs are free from hair. The stature is
tolerably equal, and is always considerably below that of
the average European; the body is robust, the breast well
developed, the feet small, thick, and short, the hands small
and rather delicate. The face is a little broad, and inclined to be flat; the forehead is rather rounded, the brows
low, the eyes black and very slightly oblique; the nose is
rather small, not prominent, but straight and well-shaped,
the apex a little rounded, the nostrils broad and slightly

exposed; the cheek-bones are rather prominent, the mouth large, the lips broad and well cut, but not protruding, the chin round and well-formed.

In this description there seems little to object to on the score of beauty, and yet on the whole the Malays are certainly not handsome. In youth, however, they are often very good-looking, and many of the boys and girls up to twelve or fifteen years of age are very pleasing, and some have countenances which are in their way almost perfect. I am inclined to think they lose much of their good looks by bad habits and irregular living. At a very early age they chew betel and tobacco almost incessantly; they suffer much want and exposure in their fishing and other excursions; their lives are often passed in alternate starvation and feasting, idleness and excessive labour,—and this naturally produces premature old age and harshness of features.

In character the Malay is impassive. He exhibits a reserve, diffidence, and even bashfulness, which is in some degree attractive, and leads the observer to think that the ferocious and bloodthirsty character imputed to the race must be grossly exaggerated. He is not demonstrative. His feelings of surprise, admiration, or fear, are never openly manifested, and are probably not strongly felt. He is slow and deliberate in speech, and circuitous in introducing the subject he has come expressly to discuss.

These are the main features of his moral nature, and exhibit themselves in every action of his life.

Children and women are timid, and scream and run at the unexpected sight of a European. In the company of men they are silent, and are generally quiet and obedient. When alone the Malay is taciturn; he neither talks nor sings to himself. When several are paddling in a canoe, they occasionally chant a monotonous and plaintive song. He is cautious of giving offence to his equals. He does not quarrel easily about money matters; dislikes asking too frequently even for payment of his just debts, and will often give them up altogether rather than quarrel with his debtor. Practical joking is utterly repugnant to his disposition; for he is particularly sensitive to breaches of etiquette, or any interference with the personal liberty of himself or another. As an example, I may mention that I have often found it very difficult to get one Malay servant to waken another. He will call as loud as he can, but will hardly touch, much less shake his comrade. I have frequently had to waken a hard sleeper myself when on a land or sea journey.

The higher classes of Malays are exceedingly polite, and have all the quiet ease and dignity of the best-bred Europeans. Yet this is compatible with a reckless cruelty and contempt of human life, which is the dark side of their character. It is not to be wondered at, therefore, that

different persons give totally opposite accounts of them—
one praising them for their soberness, civility, and good-
nature; another abusing them for their deceit, treachery,
and cruelty.   The old traveller Nicolo Conti, writing in
1430, says: "The inhabitants of Java and Sumatra exceed every other people in cruelty.   They regard killing a
man as a mere jest; nor is any punishment allotted for
such a deed.   If any one purchase a new sword, and wish
to try it, he will thrust it into the breast of the first person
he meets.   The passers-by examine the wound, and praise
the skill of the person who inflicted it, if he thrust in
the weapon direct."   Yet Drake says of the south of
Java: "The people (as are their kings) are a very loving,
true, and just-dealing people;" and Mr. Crawfurd says
that the Javanese, whom he knew thoroughly, are "a
peaceable, docile, sober, simple, and industrious people."
Barbosa, on the other hand, who saw them at Malacca
about 1660, says: "They are a people of great ingenuity,
very subtle in all their dealings; very malicious, great
deceivers, seldom speaking the truth; prepared to do all
manner of wickedness, and ready to sacrifice their lives."

The intellect of the Malay race seems rather deficient.
They are incapable of anything beyond the simplest combinations of ideas, and have little taste or energy for the
acquirement of knowledge.   Their civilization, such as it
is, does not seem to be indigenous, as it is entirely confined

to those nations who have been converted to the Mahometan or Brahminical religions.

I will now give an equally brief sketch of the other great race of the Malay Archipelago, the Papuan.

The typical Papuan race is in many respects the very opposite of the Malay, and it has hitherto been very imperfectly described. The colour of the body is a deep sooty-brown or black, sometimes approaching, but never quite equalling, the jet-black of some negro races. It varies in tint, however, more than that of the Malay, and is sometimes a dusky-brown. The hair is very peculiar, being harsh, dry, and frizzly, growing in little tufts or curls, which in youth are very short and compact, but afterwards grow out to a considerable length, forming the compact frizzled mop which is the Papuans' pride and glory. The face is adorned with a beard of the same frizzly nature as the hair of the head. The arms, legs, and breast are also more or less clothed with hair of a similar nature.

In stature the Papuan decidedly surpasses the Malay, and is perhaps equal, or even superior, to the average of Europeans. The legs are long and thin, and the hands and feet larger than in the Malays. The face is somewhat elongated, the forehead flattish, the brows very prominent; the nose is large, rather arched and high, the base thick, the nostrils broad, with the aperture hidden, owing to the

tip of the nose being elongated; the mouth is large, the
lips thick and protuberant. The face has thus an alto-

gether more European aspect than
in the Malay, owing to the large
nose; and the peculiar form of
this organ, with the more prominent brows and the character of
the hair on the head, face, and
body, enable us at a glance to
distinguish the two races. I have
observed that most of these
characteristic features are as distinctly visible in children of ten
or twelve years old as in adults,
and the peculiar form of the nose
is always shown in the figures
which they carve for ornaments

PAPUAN CHARM

to their houses, or as charms to wear round their necks.

The moral characteristics of the Papuan appear to me to
separate him as distinctly from the Malay as do his form
and features. He is impulsive and demonstrative in speech
and action. His emotions and passions express themselves
in shouts and laughter, in yells and frantic leapings.
Women and children take their share in every discussion,
and seem little alarmed at the sight of strangers and
Europeans.

Of the intellect of this race it is very difficult to judge, but I am inclined to rate it somewhat higher than that of the Malays, notwithstanding the fact that the Papuans have never yet made any advance towards civilization. It must be remembered, however, that for centuries the Malays have been influenced by Hindoo, Chinese, and Arabic immigration, whereas the Papuan race has only been subjected to the very partial and local influence of Malay traders. The Papuan has much more vital energy, which would certainly greatly assist his intellectual development. Papuan slaves show no inferiority of intellect compared with Malays, but rather the contrary ; and in the Moluccas they are often promoted to places of considerable trust. The Papuan has a greater feeling for art than the Malay. He decorates his canoe, his house, and almost every domestic utensil with elaborate carving, a habit which is rarely found among tribes of the Malay race.

In the affections and moral sentiments, on the other hand, the Papuans seem very deficient. In the treatment of their children they are often violent and cruel ; whereas the Malays are almost invariably kind and gentle, hardly ever interfering at all with their children's pursuits and amusements, and giving them perfect liberty at whatever age they wish to claim it. But these very peaceful relations between parents and children are no doubt, in a great measure, due to the listless and apathetic character of the

race, which never leads the younger members into serious
opposition to the elders; while the harsher discipline of
the Papuans may be chiefly due to that greater vigour and
energy of mind which always, sooner or later, leads to the
rebellion of the weaker against the stronger,—the people
against their rulers, the slave against his master, or the
child against its parent.

It appears, therefore, that, whether we consider their
physical conformation, their moral characteristics, or their
intellectual capacities, the Malay and Papuan races offer
remarkable differences and striking contrasts. The Malay
is of short stature, brown-skinned, straight-haired, beardless, and smooth-bodied. The Papuan is taller, is black-
skinned, frizzly-haired, bearded, and hairy-bodied. The
former is broad-faced, has a small nose, and flat eyebrows;
the latter is long-faced, has a large and prominent nose,
and projecting eyebrows. The Malay is bashful, cold,
undemonstrative, and quiet; the Papuan is bold, im-
petuous, excitable, and noisy. The former is grave and
seldom laughs; the latter is joyous and laughter-loving,—
the one conceals his emotions, the other displays them.

Having thus described in some detail, the great physical,
intellectual, and moral differences between the Malays and
Papuans, we have to consider the inhabitants of the numerous islands which do not agree very closely with either
of these races. The islands of Obi, Batchian, and the

three southern peninsulas of Gilolo, possess no true indigenous population; but the northern peninsula is inhabited by a native race, the so-called Alfuros of Sahoe and Galela. These people are quite distinct from the Malays, and almost equally so from the Papuans. They are tall and well-made, with Papuan features, and curly hair; they are bearded and hairy-limbed, but quite as light in colour as the Malays. They are an industrious and enterprising race, cultivating rice and vegetables, and indefatigable in their search after game, fish, tripang, pearls, and tortoiseshell.

In the great island of Ceram there is also an indigenous race very similar to that of Northern Gilolo. Bouru seems to contain two distinct races,—a shorter, round-faced people, with a Malay physiognomy, who may probably have come from Celebes by way of the Sula islands; and a taller bearded race, resembling that of Ceram.

Far south of the Moluccas lies the island of Timor, inhabited by tribes much nearer to the true Papuan than those of the Moluccas.

The Timorese of the interior are dusky brown or blackish, with bushy frizzled hair, and the long Papuan nose. They are of medium height, and rather slender figures. The universal dress is a long cloth twisted round the waist, the fringed ends of which hang below the knee. The people are said to be great thieves, and the tribes are always at war with each other, but they are not very courageous or bloodthirsty. The custom of " tabu," called here "pomáli," is very general, fruit trees, houses, crops, and property of all kinds being protected from depredation by this ceremony, the reverence for which is very great. A palm branch stuck across an open door, showing that the house is tabooed, is a more effectual guard against robbery than any amount of locks and bars. The houses in Timor are different from those of most of the other islands; they seem all roof, the thatch overhanging the low walls and reaching the ground, except where it is cut away for an entrance. In some parts of the west end of Timor, and on the little island of Semau, the houses more resemble those of the Hottentots, being egg-shaped, very small, and with a door only about three feet high. These are built on the ground, while those of the eastern districts are raised a few feet on posts. In their excitable disposition, loud voices, and fearless demeanour, the Timorese closely resemble the people of New Guinea.

In the islands west of Timor, as far as Flores and Sandalwood Island, a very similar race is found, which also extends eastward to Timor-laut, where the true Papuan race begins to appear. The small islands of Savu and Rotti, however, to the west of Timor, are very remarkable in possessing a different and, in some respects, peculiar race. These people are very handsome, with good features, resembling in many characteristics the race produced by the mixture of the Hindoo or Arab with the Malay. They are certainly distinct from the Timorese or Papuan races, and must be classed in the western rather than the eastern ethnological division of the Archipelago.

The whole of the great island of New Guinea, the Ké and Aru Islands, with Mysol, Salwatty, and Waigiou, are inhabited almost exclusively by the typical Papuans. I found no trace of any other tribes inhabiting the interior of New Guinea, but the coast people are in some places mixed with the browner races of the Moluccas. The same Papuan race seems to extend over the islands east of New Guinea as far as the Fijis.

There remain to be noticed the black woolly-haired races of the Philippines and the Malay peninsula, the former called "Negritos," and the latter "Semangs." I have never seen these people myself, but from the numerous accurate descriptions of them that have been published, I have had no difficulty in satisfying myself that they have little affinity or resemblance to the Papuans, with which they have been hitherto associated. In most important characters they differ more from the Papuan than they do from the Malay. They are dwarfs in stature, only averaging four feet six inches to four feet eight inches high, or eight inches less than the Malays; whereas the Papuans are decidedly taller than the Malays. The nose is invariably represented as small, flattened, or turned up at the apex, whereas the most universal character of the Papuan race is to have the nose prominent and large, with the apex produced downwards, as it is invariably represented in their own rude idols. The hair of these dwarfish races agrees with that of the Papuans, but so it does with that of the negroes of Africa. The Negritos and the Semangs agree very closely in physical characteristics with each other and with the Andaman Islanders, while they differ in a marked manner from every Papuan race.

A careful study of these varied races, comparing them with those of Eastern Asia, the Pacific Islands, and Australia, has led me to adopt a comparatively simple view as to their origin and affinities.

If we draw a line (see Physical Map, Vol. I. p. 14), commencing to the east of the Philippine Islands, thence along the western coast of Gilolo, through the island of Bouru, and curving round the west end of Flores, then bending back by Sandalwood Island to take in Rotti, we shall divide the Archipelago into two portions, the races of which have strongly marked distinctive peculiarities. This line will separate the Malayan and all the Asiatic races, from the Papuans and all that inhabit the

Pacific; and though along the line of junction intermigration and commixture have taken place, yet the division is on the whole almost as well defined and strongly contrasted, as is the corresponding zoological division of the Archipelago, into an Indo-Malayan and Austro-Malayan region.

I must briefly explain the reasons that have led me to consider this division of the Oceanic races to be a true and natural one. The Malayan race, as a whole, undoubtedly very closely resembles the East Asian populations, from Siam to Mandchouria. I was much struck with this, when in the island of Bali I saw Chinese traders who had adopted the costume of that country, and who could then hardly be distinguished from Malays ; and, on the other hand, I have seen natives of Java who, as far as physiognomy was concerned, would pass very well for Chinese. Then, again, we have the most typical of the Malayan tribes inhabiting a portion of the Asiatic continent itself, together with those great islands which, possessing the same species of large Mammalia with the adjacent parts of the continent, have in all probability formed a connected portion of Asia during the human period. The Negritos are, no doubt, quite a distinct race from the Malay ; but yet, as some of them inhabit a portion of the continent, and others the Andaman Islands in the Bay of Bengal, they must be considered to have

had, in all probability, an Asiatic rather than a Polynesian origin.

Now, turning to the eastern parts of the Archipelago, I
find, by comparing my own observations with those of the
most trustworthy travellers and missionaries, that a race
identical in all its chief features with the Papuan, is found
in all the islands as far east as the Fijis; beyond this the
brown Polynesian race, or some intermediate type, is
spread everywhere over the Pacific. The descriptions of
these latter often agree exactly with the characters of the
brown indigenes of Gilolo and Ceram.

It is to be especially remarked that the brown and
the black Polynesian races closely resemble each other.
Their features are almost identical, so that portraits of a
New Zealander or Otaheitan will often serve accurately
to represent a Papuan or Timorese, the darker colour and
more frizzly hair of the latter being the only differences.
They are both tall races. They agree in their love of art
and the style of their decorations. They are energetic,
demonstrative, joyous, and laughter-loving, and in all these
particulars they differ widely from the Malay.

I believe, therefore, that the numerous intermediate
forms that occur among the countless islands of the
Pacific, are not merely the result of a mixture of these
races, but are, to some extent, truly intermediate or transi-
tional; and that the brown and the black, the Papuan,

the natives of Gilolo and Ceram, the Fijian, the inhabitants of the Sandwich Islands and those of New Zealand, are all varying forms of one great Oceanic or Polynesian race.

It is, however, quite possible, and perhaps probable, that the brown Polynesians were originally the produce of a mixture of Malays, or some lighter coloured Mongol race with the dark Papuans ; but if so, the intermingling took place at such a remote epoch, and has been so assisted by the continued influence of physical conditions and of natural selection, leading to the preservation of a special type suited to those conditions, that it has become a fixed and stable race with no signs of mongrelism, and showing such a decided preponderance of Papuan character, that it can best be classified as a modification of the Papuan type. The occurrence of a decided Malay element in the Polynesian languages, has evidently nothing to do with any such ancient physical connexion. It is altogether a recent phenomenon, originating in the roaming habits of the chief Malay tribes ; and this is proved by the fact that we find actual modern words of the Malay and Javanese languages in use in Polynesia, so little disguised by peculiarities of pronunciation as to be easily recognisable—not mere Malay roots only to be detected by the elaborate researches of the philologist, as would certainly have been the case had their introduction been as remote·as the origin of a very·distinct race—a race as different from the Malay in mental and moral, as it is in physical characters.

As bearing upon this question it is important to point out the harmony which exists, between the line of separation of the human races of the Archipelago and that of the animal productions of the same country, which I have already so fully explained and illustrated. The dividing lines do not, it is true, exactly agree; but I think it is a remarkable fact, and something more than a mere coincidence, that they should traverse the same district and approach each other so closely as they do. If, however, I am right in my supposition that the region where the dividing line of the Indo-Malayan and Austro-Malayan regions of zoology can now be drawn, was formerly occupied by a much wider sea than at present, and if man existed on the earth at that period, we shall see good reason why the races inhabiting the Asiatic and Pacific areas should now meet and partially intermingle in the vicinity of that dividing line.

It has recently been maintained by Professor Huxley, that the Papuans are more closely allied to the negroes of Africa than to any other race. The resemblance both in physical and mental characteristics had often struck myself, but the difficulties in the way of accepting it as probable or possible, have hitherto prevented me from giving full weight to those resemblances. Geographical, zoological, and ethnological considerations render it almost certain, that if these two races ever had a common origin, it could only have been at a period far more remote than any which has yet been assigned to the antiquity of the human race. And even if their unity could be proved, it would in no way affect my argument for the close affinity of the Papuan and Polynesian races, and the radical distinctness of both from the Malay.

Polynesia is pre-eminently an area of subsidence, and its great wide-spread groups of coral-reefs mark out the position of former continents and islands. The rich and varied, yet strangely isolated productions of Australia and New Guinea, also indicate an extensive continent where such specialized forms were developed. The races of men now inhabiting these countries are, therefore, most probably the descendants of the races which inhabited these continents and islands. This is the most simple and natural supposition to make. And if we find any signs of direct affinity between the inhabitants of any other part of the world and those of Polynesia, it by no means follows that the latter were derived from the former. For as, when a Pacific continent existed, the whole geography of the earth's surface would probably be very different from what it now is, the present continents may not then have risen above the ocean, and, when they were formed at a subsequent epoch, may have derived some of their inhabitants from the Polynesian area itself. It is undoubtedly true that there are proofs of extensive migrations among the Pacific islands, which have led to community of language from the Sandwich group to New Zealand; but there are no proofs whatever of recent migration from any surrounding country to Polynesia, since there is no people to be found elsewhere sufficiently resembling the Polynesian race in their chief physical and mental characteristics.

If the past history of these varied races is obscure and uncertain, the future is no less so. The true Polynesians, inhabiting the farthest isles of the Pacific, are no doubt doomed to an early extinction. But the more numerous Malay race seems well adapted to survive as the cultivator of the soil, even when his country and government have passed into the hands of Europeans. If the tide of colonization should be turned to New Guinea, there can be little doubt of the early extinction of the Papuan race. A warlike and energetic people, who will not submit to national slavery or to domestic servitude, must disappear before the white man as surely as do the wolf and the tiger.

I have now concluded my task. I have given, in more or less detail, a sketch of my eight years' wanderings among the largest and the most luxuriant islands which adorn our earth's surface. I have endeavoured to convey my impressions of their scenery, their vegetation, their animal productions, and their human inhabitants. I have dwelt at some length on the varied and interesting problems they offer to the student of nature. Before bidding my readers farewell, I wish to make a few observations on a subject of yet higher interest and deeper importance, which the contemplation of savage life has suggested, and on which I believe that the civilized can learn something from the savage man.

We most of us believe that we, the higher races, have progressed and are progressing. If so, there must be some state of perfection, some ultimate goal, which we may never reach, but to which all true progress must bring us nearer. What is this ideally perfect social state towards which mankind ever has been, and still is tending ? Our best thinkers maintain, that it is a state of individual freedom and self-government, rendered possible by the equal development and just balance of the intellectual, moral and physical parts of our nature,—a state in which we shall each be so perfectly fitted for a social existence, by knowing what is right, and at the same time feeling an irresistible impulse to do what we know to be right, that all laws and all punishments shall be unnecessary In such a state every man would have a sufficiently well-balanced intellectual organization, to understand the moral law in all its details, and would require no other motive but the free impulses of his own nature to obey that law.

Now it is very remarkable, that among people in a very low stage of civilization, we find some approach to such a perfect social state. I have lived with communities of savages in South America and in the East, who have no laws or law courts but the public opinion of the village freely expressed. Each man scrupulously respects the rights of his fellow, and any infraction of those rights rarely or never takes place. In such a community, all are nearly equal. There are none of those wide distinctions, of education and ignorance, wealth and poverty, master and servant, which are the product of our civilization; there is none of that wide-spread division of labour, which, while it increases wealth, produces also conflicting interests; there is not that severe competition and struggle for existence, or for wealth, which the dense population of civilized countries inevitably creates. All incitements to great crimes are thus wanting, and petty ones are repressed, partly by the influence of public opinion, but chiefly by that natural sense of justice and of his neighbour's right, which seems to be, in some degree, inherent in every race of man.

Now, although we have progressed vastly beyond the savage state in intellectual achievements, we have not advanced equally in morals. It is true that among those classes who have no wants that cannot be easily supplied, and among whom public opinion has great influence, the rights of others are fully respected. It is true, also, that we have vastly extended the sphere of those rights, and include within them all the brotherhood of man. But it is not too much to say, that the mass of our populations have not at all advanced beyond the savage code of morals, and have in many cases sunk below it. A deficient morality is the great blot of modern civilization, and the greatest hindrance to true progress.

During the last century, and especially in the last thirty years, our intellectual and material advancement has been too quickly achieved for us to reap the full benefit of it. Our mastery over the forces of nature has led to a rapid growth of population, and a vast accumulation of wealth; but these have brought with them such an amount of poverty and crime, and have fostered the growth of so much sordid feeling and so many fierce passions, that it may well be questioned, whether the mental and moral status of our population has not on the average been lowered, and whether the evil has not overbalanced the good. Compared with our wondrous progress in physical science and its practical applications, our system of government, of administering justice, of national education, and our whole social and moral organization, remains in a state of barbarism.* And if we continue to devote our chief energies to the utilizing of our knowledge of the laws of nature with the view of still further extending our commerce and our wealth, the evils which necessarily accompany these when too eagerly pursued, may increase to such gigantic dimensions as to be beyond our power to alleviate.

We should now clearly recognise the fact, that the wealth and knowledge and culture of *the few* do not constitute civilization, and do not of themselves advance us towards the " perfect social state." Our vast manufacturing system, our gigantic commerce, our crowded towns and cities, support and continually renew a mass of human misery and crime *absolutely* greater than has ever existed before. They create and maintain in life-long labour an ever-increasing army, whose lot is the more hard to bear, by contrast with the pleasures, the comforts, and the luxury which they see everywhere around them, but which they can never hope to enjoy; and who, in this respect, are worse off than the savage in the midst of his tribe.

This is not a result to boast of, or to be satisfied with ; and, until there is a more general recognition of this failure of our civilization—resulting mainly from our neglect to train and develop more thoroughly the sympathetic feelings and moral faculties of our nature, and to allow them

* See note next page.

a larger share of influence in our legislation, our commerce, and our whole social organization—we shall never, as regards the whole community, attain to any real or important superiority over the better class of savages.

This is the lesson I have been taught by my observations of uncivilized man. I now bid my readers— Farewell!

NOTE.

THOSE who believe that our social condition approaches perfection, will think the above word harsh and exaggerated, but it seems to me the only word that can be truly applied to us. We are the richest country in the world, and yet one-twentieth of our population are parish paupers, and one-thirtieth known criminals. Add to these, the criminals who escape detection, and the poor who live mainly on private charity, (which, according to Dr. Hawkesley, expends seven millions sterling annually in London alone,) and we may be sure that more than ONE-TENTH of our population are actually Paupers and Criminals. Both these classes we keep idle or at unproductive labour, and each criminal costs us annually in our prisons more than the wages of an honest agricultural labourer. We allow over a hundred thousand persons known to have no means of subsistence but by crime, to remain at large and prey upon the community, and many thousand children to grow up before our eyes in ignorance and vice, to supply trained criminals for the next generation. This, in a country which boasts of its rapid increase in wealth, of its enormous commerce and gigantic manufactures, of its

mechanical skill and scientific knowledge, of its high civilization
and its pure Christianity,—I can but term a state of social
barbarism. We also boast of our love of justice, and that the
law protects rich and poor alike, yet we retain money fines as a
punishment, and make the very first steps to obtain justice a
matter of expense—in both cases a barbarous injustice, or denial
of justice to the poor. Again, our laws render it possible, that,
by mere neglect of a legal form, and contrary to his own wish
and intention, a man's property may all go to a stranger, and
his own children be left destitute. Such cases have happened
through the operation of the laws of inheritance of landed property ; and that such unnatural injustice is possible among us,
shows that we are in a state of social barbarism. One more
example to justify my use of the term, and I have done. We
permit absolute possession of the soil of our country, with no
legal rights of existence on the soil, to the vast majority who do
not possess it. A great landholder may legally convert his whole
property into a forest or a hunting-ground, and expel every
human being who has hitherto lived upon it. In a thickly-
populated country like England, where every acre has its owner
and its occupier, this is a power of legally destroying his fellow-
creatures ; and that such a power should exist, and be exercised
by individuals, in however small a degree, indicates that, as
regards true social science, we are still in a state of barbarism.

APPENDIX.

APPENDIX.

ON THE CRANIA AND THE LANGUAGES OF THE RACES OF MAN IN THE MALAY ARCHIPELAGO.

CRANIA.

A FEW years ago it was thought that the study of Crania offered the only sure basis of a classification of man. Immense collections have been formed; they have been measured, described, and figured; and now the opinion is beginning to gain ground, that for this special purpose they are of very little value. Professor Huxley has boldly stated his views to this effect; and in a proposed new classification of mankind has given scarcely any weight to characters derived from the cranium. It is certain, too, that though Cranioscopy has been assiduously studied for many years, it has produced no results at all comparable with the labour and research bestowed upon it. No approach to a theory of the excessive variations of the cranium has been put forth, and no intelligible classification of races has been founded upon it.

Dr. Joseph Barnard Davis, who has assiduously collected human crania for many years, has just published a remarkable work, entitled " Thesaurus Craniorum." This is a catalogue of his collection (by far the most extensive in existence), classified according to countries and races, indicating the derivation and any special characteristics of each specimen; and by way of description, an elaborate series of measurements, nineteen in number when complete, by which accurate comparisons can be made, and the limits of variation determined.

This interesting and valuable work offered me the means of determining for myself, whether the forms and dimensions of the crania of the eastern races, would in any way support or refute my classification of them. For the purposes of comparison, the whole series of nineteen measurements would have been far too cumbersome. I therefore selected three, which seem to me well adapted to test the capabilities of Cranioscopy for the purpose in view. These are :—1. The capacity of the cranium. 2. The proportion of the width to the length taken as 100. 3. The proportion of the height to the length taken as 100. These dimensions are given by Dr. Davis in almost every case, and have furnished me with ample materials. I first took the "*means*" of groups of crania of the same race from distinct localities, as given by Dr. Davis himself, and thought I could detect differences characteristic of the great divisions of the Malayans and Papuans; but some anomalies induced me to look at the amount of individual variation, and this was so enormous that I became at once convinced, that even this large collection could furnish no trustworthy average. I will now give a few examples of these variations, using the terms,—Capacity, W:L, H:L, for the three dimensions compared. In the Capacity, I always compare only male crania, so as not to introduce the sexual difference of size. In the other proportionate dimensions, I use both sexes to get a larger average, as I find these proportions do not vary definitely according to sex, the two extremes often occurring in the series of male specimens only.

MALAYS.—Thirteen male Sumatra crania had :—Capacity, from 61·5 to 87 ounces of sand; W:L, ·71 to ·86; H:L, 73 to ·85. Ten male Celebes crania varied thus :—Capacity, from 67 to 83; W : L, ·73 to ·92; H:L, ·76 to ·90.

In the whole series of eighty-six Malay skulls from Sumatra, Java, Madura, Borneo, and Celebes, the variation is enormous. Capacity (66 skulls) 60 to 91 ounces of sand; W:L, ·70 to 92; H : L, ·72 to ·90. And these extremes are not isolated abnormal specimens, but there is a regular gradation up to them, which always becomes more perfect the larger the number of specimens compared. Thus, besides the extreme Dolicocephalic skull (70) in the supposed Brachycephalic Malay group, there are others which have W : L, ·71, ·72 and ·73, so that we have every reason to believe that with more specimens we should get a still narrower form of skull. So the very large cranium, 91 ounces, is led up to by others of 87 and 88.

The largest, in an extensive series of English, Scotch, and Irish crania, was only 92·5 ounces.

PAPUANS.—There are only four true Papuan crania in the collection, and these vary considerably (W : L ·72 to ·83). Taking, however, the natives of the Solomon Islands, New Caledonia, New Hebrides, and the Fijis as being all decidedly of Papuan race, we have a series of 28 crania (23 male), and these give us :—Capacity, 66 to 80 ; W:L, ·65 to ·85 ; H:L, 71 to ·85 ; so nearly identical with some of the Malayan groups as to offer no clear points of difference.

The Polynesians, the Australians, and the African negroes offer equally wide ranges of variation, as will be seen by the following summary of the dimensions of the crania of these races and the preceding :—

Number of Crania.	CAPACITY	W L	H L
83. Malays (66 male).	60 to 91	·70 to ·92	·72 to 90
28. Papuans (23 m.) .	66 ,, 80	·65 ,, ·85	·71 ,, ·85
156. Polynesians (90 m.)	62 ,, 91	·69 ,, 90	·68 ,, ·88
23. Australians (16 m.)	59 ,, 86	·57 ,, ·80	·64 ,, ·80
72. Negroes (38 m) .	66 ,, 87	·64 ,, ·83	·65 ,, ·81

The only conclusions that we can draw from this table are, that the Australians have the smallest crania, and the Polynesians the largest; the Negroes, the Malays, and Papuans not

differing perceptibly in size.   And this accords very well
with what we know of their mental activity and capacity for
civilization.

The Australians have the *longest* skulls ; after which come the
Negroes ; then the Papuans, the Polynesians, and the Malays.

The Australians have also the *lowest* skulls ; then the Negroes ;
the Polynesians and Papuans considerably higher and equal, and
the Malay the highest.

It seems probable, therefore, that if we had a much more extensive series of crania the averages might furnish tolerably reliable
race-characters, although, owing to the large amount of individual
variation, they would never be of any use in single examples,
or even when moderate numbers only could be compared.

So far as this series goes, it seems to agree well with the
conclusions I have arrived at, from physical and mental characters observed by myself.   These conclusions briefly are : that
the Malays and Papuans are radically distinct races ; and that
the Polynesians are most nearly allied to the latter, although they
have probably some admixture of Malayan or Mongolian blood.

LANGUAGES.

During my travels among the islands of the Archipelago, I collected a considerable number of vocabularies, in districts hitherto
little visited. These represent about fifty-seven distinct languages
(not including the common Malay and Javanese), more than half
of which I believe are quite unknown to philologists, while only
a few scattered words have been recorded of some others.
Unfortunately, nearly half the number have been lost.   Some
years ago I lent the whole series to the late Mr. John Crawford,
and having neglected to apply for them for some months, I
found that he had in the meantime changed his residence, and
that the books, containing twenty-five of the vocabularies, had
been mislaid ; and they have never since been recovered. Being
merely old and much battered copy-books, they probably found

their way to the dust-heap along with other waste paper. I had previously copied out nine common words in the whole series of languages, and these are here given, as well as the remaining thirty-one vocabularies in full.

Having before had experience of the difficulty of satisfactorily determining any words but nouns and a few of the commonest adjectives, where the people are complete savages and the language of communication but imperfectly known, I selected about a hundred and twenty words, and have adhered to them throughout as far as practicable. After the English, I give the Malay word for comparison with the other languages. In orthography I adopt generally the continental mode of sounding the vowels, with a few modifications, thus :—

English . . . . a e i *or* ie ei o ŭ   ū
Sounded. . . . ah a ee   i o é *or* éh oo

These sounds come out most prominently at the end of a syllable; when followed by a consonant the sounds are very little different from the usual pronunciation. Thus, "Api" is pronounced *Appee*, while "Minta" is pronounced *Mintah.* The short ŭ is pronounced like *er* in English, but without any trace of the guttural. Long, short, and accented syllables are marked in the usual way. The languages are grouped geographically, passing from west to east ; those from the same or adjacent islands being as much as possible kept together.

I profess to be able to draw very few conclusions from these vocabularies. I believe that the languages have been so much modified by long intercommunication among the islands, that resemblances of words are no proof of affinity of the people who use those words. Many of the wide-spread similarities can be traced to organic onomatopœia. Such are the prevalence of *g* (hard), *ng, ni,* in words meaning "tooth;" of *l* and *m* in those for "tongue ;" of *nge , ung, sno,* in those for "nose." Others are plainly commercial words, as " salaka " and

" ringgit " (the Malay word for dollar) for silver, and " mas "
for gold. The Papuan group of languages appear to be distinguished by harsher combinations of letters, and by monosyllabic
words ending in a consonant, which are rarely or never found in
the Malay group. Some of the tribes who are decidedly of
Malay race, as the people of Ternate, Tidore, and Batchian, speak
languages which are as decidedly of a Papuan type ; and this,
I believe, arises from their having originally immigrated to these
islands in small numbers, and by marrying native women acquired
a considerable portion of their language, which later arrivals of
Malays were obliged to learn and adopt if they settled in the
country. As I have hardly mentioned in my narrative some of
the names of the tribes whose languages are here given, I will
now give a list of them, with such explanatory remarks as I may
think useful to the ethnologist, and then leave the vocabularies
to speak for themselves.

LIST OF VOCABULARIES COLLECTED

*Those marked * are lost.*

1. **Malay.**— The common colloquial Malay as spoken in
Singapore ; written in the Arabic character.

2. **Javanese.**— Low or colloquial Javanese as spoken in
Java; written in a native character.

*3. **Sassak.**—Spoken by the indigenes of Lombock, who are
Mahometans, and of a pure Malay race.

*4. **Macassar.**—Spoken in the district of Southern Celebes,
near Macassar ; written in a native character.   Mahometans.

*5. **Bugis.**—Spoken over a large part of Southern Celebes ;
written in a native character distinct from that of Macassar.
Mahometans.

6. **Bouton.**—Spoken in Boutong, a large island south of
Celebes.   Mahometans.

7. **Salayer.**—Spoken in Salayer, a smaller island south of Celebes. Mahometans.

*8. **Tomore.**—Spoken in the eastern peninsula of Celebes, and in Batchian, by emigrants who have settled there. Pagans.

Note.—The people who speak these five languages of Celebes are of pure Malayan type, and (all but the last) are equal in civilization to the true Malays.

*9. **Tomohon**; *10. **Langowen.**—Villages on the plateau of Minahasa.

*11. **Ratahan**; *12. **Belang.**—Villages near the south-east coast of Minahasa. *13. **Tanawanko.**—On the west coast. *14. **Kema.**—On the east coast. *15. **Bantek.**—A suburb of Menado.

16. **Menado.**—The chief town. 17. **Bolang-hitam.**—A village on the north-west coast, between Menado and Licoupang.

These nine languages, with many others, are spoken in the north-west peninsula of Celebes, by the people called Alfuros, who are of Malay race, and seem to have affinities with the Tagalas of the Philippines through the Sanguir islanders. These languages are falling into disuse, and Malay is becoming the universal means of communication. Most of the people are being converted to Christianity.

18. **Sanguir Islands** and **Siau.**—Two groups of islands between Celebes and the Philippines. The inhabitants wear a peculiar costume, consisting of a loose cotton gown hanging from the neck nearly to the feet. They resemble, physically, the people of Menado.

19. **Salibabo Islands,** also called **Talaut.**—This vocabulary was given me from memory by Captain Vanderbeck. See page 76.

20. **Sula Islands.**—These are situated east of Celebes, and their inhabitants seem to be Malays of the Moluccan type, and are Mahometans.

21. Ca¦eli; 22. **Wayapo**; 23. **Massaratty.**—These are

three villages on the eastern side of Bouru. The people are allied to the natives of Ceram. Those of Cajeli itself are Mahometans.

24. **Amblau.** — An island a little south-east of Bouru. Mahometans.

*25 **Ternate.**—The northernmost island of the Moluccas. The inhabitants are Mahometans of Malay race, but somewhat mixed with the indigenes of Gilolo.

26. **Tidore.**—The next island of the Moluccas. The inhabitants are undistinguishable from those of Ternate.

*27. **Kaióa Islands.**—A small group north of Batchian.

*28. **Batchian.**—Inhabitants like the preceding. Mahometans, and of a similar Malay type.

29. **Gani.**—A village on the south peninsula of Gilolo. Inhabitants, Moluccan-Malays, and Mahometans.

*30. **Sahoe;** 31. **Galela.**—Villages of Northern Gilolo. The inhabitants are called Alfuros. They are indigenes of Polynesian type, with brown skins, but Papuan hair and features Pagans.

32. **Liang.** — A village on the north coast of Amboyna. Several other villages near speak the same language. They are Mahometans or Christians, and seem to be of mixed Malay and Polynesian type.

33. **Morella** and **Mamalla.** — Villages in North-West Amboyna. The inhabitants are Mahometans.

34. **Batu-merah.** — A suburb of Amboyna. Inhabitants Mahometans, and of Moluccan-Malay type.

35. **Lariki, Asilulu, Wakasiho.**—Villages in West Amboyna inhabited by Mahometans, who are reported to have come originally from Ternate.

36. **Saparua.**—An island east of Amboyna. Inhabitants of the brown Polynesian type, and speaking the same language as those on the coast of Ceram opposite.

37. **Awaiya;** 38. **Camarian.**—Villages on the south coast of Ceram. Inhabitants indigenes of Polynesian type, now Christians.

39. **Teluti** and **Hoya**; 40. **Ahtiago** and **Tobo.**—Villages on the south coast of Ceram. Inhabitants Mahometans, of mixed brown Papuan or Polynesian and Malay type.

41. **Ahtiago.**—Alfuros or indigenes inland from this village. Pagans, of Polynesian or brown Papuan type.

42. **Gah.**—Alfuros of East Ceram.

43. **Wahai.**—Inhabitants of much of the north coast of Ceram. Mahometans of mixed race. Speak several dialects of this language.

*44. **Goram.**—Small islands east of Ceram. Inhabitants of mixed race, and Mahometans.

45. **Matabello.**—Small islands south-east of Goram. Inhabitants of brown Papuan or Polynesian type. Pagans.

46. **Teor.**—A small island south-east of Matabello. Inhabitants a tall race of brown Papuans. Pagans.

*47. **Ké Islands.**—A small group west of the Aru Islands. Inhabitants true black Papuans. Pagans.

*48. **Aru Islands.**—A group west of New Guinea. Inhabitants true Papuans. Pagans.

49. **Mysol** (coast).—An island north of Ceram. Inhabitants Papuans with mixture of Moluccan Malays. Semi-civilized.

50. **Mysol** (interior).—Inhabitants true Papuans. Savages.

*51. **Dorey.**—North coast of New Guinea. Inhabitants true Papuans. Pagans.

*52. **Teto**; *53. **Vaiqueno, East Timor**; *54. **Brissi, West Timor.**—Inhabitants somewhat intermediate between the true and the brown Papuans. Pagans.

*55. **Savu**; *56. **Rotti.**—Islands west of Timor. Inhabitants of mixed race, with apparently much of the Hindoo type.

*57. **Allor**; *58. **Solor.**—Islands between Flores and Timor. Inhabitants of dark Papuan type.

59. **Bajau,** or **Sea Gipsies.**—A roaming tribe of fishermen of Malayan type, to be met with in all parts of the Archipelago.

Nine Words in Fifty-nine Languages

English.	BLACK.	FIRE.	LARGE.	NOSE.
1. Malay	Itam	Api	Búsar	Idong
2 Javanese	Iran	Gūni	Gedé	Irong
3. Sasak (Lombock)	Bidan	Api	Ble	Idong
4. Macassar	Leling	Pepi	Lompo	Kamūrong
5. Bugis	Malotong	Api	Marája	Ingok
6. Bouton	Amáita	Whá	Monghi	Oánu
7. Salayer	Hitam	Api	Bakeh	Kumor
8. Tomóre	Moito	Api	Owhosi	Hengénto
9. Tomohon	Rūmdum	Api	Tuwón	Ngerun
10. Langowan	Wūlin	Api	Wanko	Ngilung
11. Ratahan	Mahítum	Pūtong	Loben	Irun
12. Belang	Mūhónde	Sūlu	Musolah	Niyun
13. Tanawanko	Rūmdum	Api	Sūla	Ngerun
14. Kema	Hirun	Api	Sūla	Ngerun
15. Bantek	Maitung	Pūtung	Ramoh	Idung
16. Menado	Maitung	Pūtung	Raboh	Idong
17. Bolang Itam	Moitomo	Pūro	Morokaro	Djunga
18. Sanguir Is.	Maitum	Pūtun	Labo	Hirong
19. Salibabo Is.	Maitu	Pūton	Bagewa	
20. Sula Is.	Miti	Api	Ea	Ne
21. Cajeli	Metan	Ahú	Lchai	Nem
22. Wayapo	Miti	Bána	Bagut'	Nien
23. Massaratty	Miti	Bana	Haat	Nieni
24. Amblaw	Kameichei	Afu	Plaré	Neinya téha
25. Ternate	Kokotu	Uku	Lamu lamu	Nunu
26. Tidore	Kokótu	Uku	Lamu	Un
27. Kaióa Is.	Kūda	Lūtan	Lol	Usnod
28. Batchian	Ngóa	Api	Rá	Hidom
29. Gani	Kitkudu	Lūtan	Talalólo	Usnut
30 Sahoe	Kokótu	Uhuh	Lamu	Ngūnu
31 Galela	Tatataio	Uku	Elamo.	Ngūno
32. Liang	Méte	Aów	Nila	Hiruka
33. Morella	Méte	Aów	Hella	Iuka
34. Batu-merah	Meteni	Aow	Enda-á	Ninura
35. Lariki, &c	Méte	Aow	Era	Iru
36. Saparua	Meteh	Háo	Ilahil	Iri
37 Awaiya	Meténi	Aousa	Ilahe	Nua-mo
38. Camarian	Méti	Hao	Eraámei	Hili-mo
39. Teluti	Méte	Yafo	Elau	Olicolo
40. Ahtiago (Mah)	Memétan	Yaf	Aiyuk	Ilin
41. Ahtiago (Alf.)	Meten	Wahum	Poten	Ilnum
42. Gah	Miatan	Aif	Boluk	Sonina
43. Wahai	Meten	Aow	Maina	Inóre
44. Goram	Meta metan	Hai	Bobok	Suwera
45. Matabello	Moten	Efi	Leleh	Wiramáni
46. Teor	Miten	Yaf	Lēn	Gilinkani
47. Ké Is.	Metan	Youf	Lih	Nirun
48. Aru Is.	Būré	Ow	Jinny	Djurul
49. Mysol (Coast)	Mūlmetan	Lap	Sala	Shong gulu
50. Do. (Interior)	Bit	Yap	Klen	Mot mobi
51. Dorey	Paisin	Voor	Iba	Snori'
52. Teto, E.	Metan	Hahi	Bot	Inur
53. Vaiqueno, E.	Meta	Hai	Naiki	Inu
54. Brissi, W.	Metan	Ai	Naaik, Bena	Panan
55. Savu	Meddi	Ai	Moneái	Hewonga
56. Rotti	Ngéo	Hai	Matua, Malóa	Idun
57. Allor	Mité	Api	Bé	Niru
58. Solor	Mitang	Api	Belang	Irung
59. Bájau (Sea Gipsies)	Lawon	Api	Basar	Uroh

Rows 4–8 bracketed: S. Celebes.
Rows 9–17 bracketed: North Celebes.
Rows 21–24 bracketed: Bouru.
Rows 27–31 bracketed: Gilolo.
Rows 32–43 bracketed: Amboyna. / Ceram.
Rows 52–54 bracketed: Timor.

OF THE MALAY ARCHIPELAGO.

	SMALL.	TONGUE.	TOOTH.	WATER.	WHITE.
1.	Kíchil	Lídah	Gígi	Ayer	Pūtih.
2.	Chili	Ilat	Untu	Banyu	Pūteh.
3.	Bri	Ellah	Gigi	Aie	Pūtih.
4.	Chadi	Lelah	Gigi	Yéni	Kebo.
5.	Becho	Lila	Isi	Uwál	Mapūte.
6.	Kidikidi	Lilah	Nichi	Mánu	Mapūti.
7.	Kedi	Lilah	Gígi	Aer	Pūtih
8.	Odidi	Elunto	Nisinto	Mánu	Mopūtih.
9.	Koki	Lilah	Baan	Rano	Kuloh.
10.	Toyáan	Lilah	Ipau	Rano	Kuloh.
11.	Iok	Rilah	Isi	Aki	Mawuroh.
12.	Mohintek	Lilah	Mopon	Tivi	Pūtih.
13.	Koki	Lilah	Waán	Rano	Kūloh.
14.	Koki	Dilah	Waang	Dorr	Pūtih.
15.	Kokonio	Dilrah	Isy	Akéi	Mabida.
16.	Dodío	Lilah	Ngisi	Akéi	Mabida.
17.	Moisiko	Dila	Dongito	Sarugo	Mopótiho.
18.	Anióu	Lilah	Isi	Aki	Mawérah.
19.	Kadodo			Wai	Mawirah.
20.	Mahé	Maki	Nihi	Wai	Bóti.
21.	Koi	Mahino	Nisini	Waili	Umpoti.
22.	Roit	Maan	Nisi	Wai	Boti.
23.	Roi	Maanen	Nisinen	Wai	Boti.
24.	Bakoti	Munartea	Nisnya-teha	Wai	Purini.
25.	Ichi ichi	Aki	Ingin	Namo	Bobūdo.
26.	Kéni	Aki	Ing	Aki	Bubūlo
27.	Kūtu	Mod	Hahlo	Woya	Bulam.
28.	Díkit	Lidah	Gigi	Paisu	Putih.
29.	Wai-waio	Imōd	Afod	Waiyr.	Wūlan.
30.	Cheka	Yeidi	Ngedi	Namo	Būdo.
31.	Dechéki	Nangaládi	Ini	Aki	Daari.
32.	Koi	Meka	Niki	Wehr	Pūtih.
33.	Ahuntai	Meka	Nikin	Wehl	Pūtih.
34.	Ana-á	Numawa	Nindíwa	Weyl	Pūtih.
35.	Koi	Méh	Niki	Weyl	Pūtih.
36.	Ihihil	Mé	Nio	Wai	Pūtil.
37.	Olihil	Mei	Nisi-mo	Waeli	Pūtile.
38.	Kokanéii	Meem	Nikim	Waeli	Pūtih.
39.	Anan	Mecolo	Lilico	Welo	Pūtih.
40.	Nelak	Melin	Nifan	Wai	Babut
41.	Anaanin	Ninúm	Nesnim	Waiin	Pūtih.
42.	Wota wota	Lemukonína	Nisikonina	Arr	Maplutu.
43.	Kiiti	Mé	Lesin	Tólun	Pūteh.
44.	Tutúin	Kelo	Nisium	Arr	Mehūti.
45.	Enéna	Tumoma	Nifoa	Arr	Maphūti
46.	Fek	Mēn	Nifin	Wehr	Sélūp.
47.	Kot	Nefan	Oin	Wehr	Neah.
48.	Sie	Gigi	Mulu	Wehr	Eren.
49.	Gūnain	Aran	Kalifin	Wayr	Būs.
50.	Senpoh	Aran	Kelif		Boo.
51.	Besarbamba	Kaprendi	Nasi	Waar	Piūper.
52.	Luik	Nañal	Nian	Vé	Mūty.
53.	Anâ	Icmal	Nissy	Hoi	Mūty.
54.	Ana	Man	Nissin	Ou	Mūty.
55.	Anaíki	Weo	Ngútu	Uilóko	Pūdi.
56.	Anoána, Loaána.	Máan	Nissi	Oée	Fūla.
57.	Kaái	Wewelli	Ulo	Wé	Būráka.
58.		Ewel	Ipa	Wai	Būrang.
59.	Didiki	Délah	Gígi	Boi	Potih.

One Hundred and Seventeen Words in Thirty-three

English	ANT.	ASHES.	BAD.	BANANA.
1. Malay	Sŭmut	Hábū	Jáhat	l'ísang
2. Javanese	Sūmut	A'vu	Ollo	Gudang
6. Bouton } S. Celebes.	Oséa	Orápu	Madúki	Olóka
7. Salayer }	Kalihara	Umbo	Seki	Loka
16. Menado } N. Celebes	Singeh	Abū	Dalruy	Lénsa
17. Bolang-hitam }	Tohomo	Awu	Moiatu	Pagie
18. Sanguir, Sian	Kiáso	Henáni	Lai	Busa
19. Salibabo			Reoh	
20. Sula Is.	Kokoi	Aftúha	Busár	Fía
21. Cajeli } Bouru.	Mosisin	Aptai	Nakié	Umpúlue
22. Wayapo }	Fosisin	Aptai	Dabóho	Fūat
23. Massaratty }	Misisin	Ogotĭn	Dabóho	Fúati
24. Amblaw	Kakai	Lávu	Behei	Biyeh
26. Tidore	Bifi	Fíka	Jíra	Koi
29. Gani } Gilolo.	Laim	Tapin	Lekat	Lókka
31. Galela }	Golúdo	Kapok	Atoró	Bóle
32. Liang } Amboyna.	Umu	Awmáti	Ahia	Kula
33. Morella }	Oön	Armatei	Ahia	Kula
34. Batumerah }	Manisiá	Howaluxi	Akahia	Iáni
35. Lariki }	Aten	Aow matei	Ahia	Kōra
36. Saparua	Sumakow	Hamatanyo	Ahía	Kúla
37. Awaiya } Ceram.	Tumúie	Ahwotoí	Ahia	Wūri
38. Camarian }	Sümukáo	Hao matei	Ahié	U'ki
39. Teluti }	Phóino	Yafow matán.	Ahia	Peléwa
40. Ahtiago and Tobo }	Fóin	Laftaín	A'vet	Fūd
41. Ahtiago (Alfuros) }		Laf teinim	Kafetáia	Phitim
42. Gah }	Niéfer	Aif tai	Nungalótuk	Fúdia
43. Wahai }	Isalema	Tókar	Aháti	Uri
45. Matabello	Otúma	Aow lómi	Ráhat	Phúdi⁻
46. Teor	Singa singat.	Yaf leit	Yat	Mūk
49. Mysol	Kamili	Gelap	Lek	Talah
50. Mysol	Kumlih	Geni	Leak	Máh
59. Baju	Sumut	Habu	Ráhat	Pisang

LANGUAGES OF THE MALAY ARCHIPELAGO.

	BELLY.	BIRD.	BLACK.	BLOOD.	BLUE.	BOAT.
1.	Prút	Bŭrung	Itam	Dárah	Bíru	Praŭ.
2.	Wŭtan	Manok	Iran	Gŭte	Biru	Prau.
6.	Kompo	Manumanu	Amaíta	Oráh	Ijan	Búnka.
7.	Pompon	Burung	Hitam	Rara	Láo	Lopi.
16.	Tïjan	Mánu	Maitung	Daha	Mabidu	Sakaen.
17.	Teo	Manoko	Moitomo	Dugu	Morono	Bolato.
18.	Tian	Manu	Maitun	Daha	Biru	Sakaen.
19.		Manu urarutang	Ma-itu		Biru	Kasáneh.
20.	Téna	Mánu	Miti	Póha	Biru	Lótu.
21.	Tihumo	Manúi	Métan	Lála	Biru	Waä.
22.	Tihen	Manúti	Miti	Raha	Biru	Wága.
23.	Fukanen	Mánúti	Miti	Ráha	Biru	Waga.
24.	Remnati kuroi	Manúe	Kame ichei	Hahanatéa	Biroi	Waä.
26.	Yóru	Namo bangow	Kokótu	Yán	Rúru	O'ti.
29.	Tutut	Manik	Kitkúdu	Sislor	Biru	Wōg.
31.	Poko	Namo	Tatatáro	Larahnangow	Biru	Déru.
32.	Hatuáka	Tuwi	Méte	Lala	Mala	Haka.
33.	Tiáka	Mano	Méte	Lala	Mala	Haka.
34.	Tiáva	Burung	Meténi	Lalai	Amála	Háka.
35.	Tia	Mano	Méte	Lala	Mála	Sepó.
36.	Teho	Mano	Moteh	Lalah	Lala	Tala.
37.	Tia	Manúe	Meténi	Lalah	Meteni	Siko.
38.	Tiámo	Mánu	Méti	Lála	Lála	Tála.
39.	Teocólo	Manúo	Méte	Láia	Lala	Yalopeí.
40.	Tian	Nióva	Memétan	Láwa	Biru	Wáha.
41.	Tapura	Manuwan	Meten	Lahim	Masounanini	Waim.
42.	Toniña	Manok	Miatan	Lalai	Biri	Wúna.
43.	Tiare	Malok	Meten	Lasin	Marah	Folútu.
45.	Abúda	Mánok	Meten	Lárah	Biru	Sóa.
46.	Kabin	Manok	Miten	Larah	Biru	Hól.
49.	Nan		Mulmetan	Lomos	Melah	Owé.
50.	Mot ni		Bít	Lemoh		Owáwi.
59.	Bútah	Mano	Lawön	Lahah	Lawu	Bido.

One Hundred and Seventeen Words in Thirty-three

English	BODY.	BONE.	BOW.	BOX.
1. Malay	Bádan	Túlang	Pánah	Púti
2. Javanese	Awah	Bálong	Panah	Krobak
6. Bouton } S. Celebes	Karóko	Obúku	Opána	Buéti
7. Salayer }	Kaleh	Boko	Panah	Puti
16. Menado } N. Celebes	Dokoku, Aoh.	Duhy		Mabida
17. Bolang- }				
hitam }	Botanga	Tula		
18. Sanguir, Sian.	Badan	Buko		Bantali
19. Salibabo			Papite	
20. Sula Is.	Kóli	Hoi	Djūb	Burúa
21. Cajeli	Batum	Lolimo	Panah	Bueti
22. Wayapo } Bouru.	Fatan	Rohin		Buéti
23. Massaratty... }	Fatanin	Rohin	Pánat	Buéti
24. Amblaw	Nanau	Koknatéa	Busu	Poroso
26. Tidore	Róhi	Yóbo	Jobi johi	Barúa
29. Gani } Gilolo.	Badan	Momud	Pusi	Barúa
31. Galela }	Nangaróhi	Kovo	Ngámi	Barúa
32. Liang } Amboyna.	Nanáka	Ruri	Husur	Buéti
33. Morella }	Dada	Luli	Husul	Buéti
34. Batumerah }	Anáro	Lulivá	Apúsu	Saüpa
35. Lariki }	Anána	Ruri	Husur	Buéti
36. Saparua.	Inawallah	Riri	Husu	Ruūwai
37. Awaiya } Ceram.	Sanawála	Lila	Husúli	Pūéti
38. Camarian }	Patani	Nili	Husúli	Buéti
39. Teluti }	Hatáko	Toicólo	Osio	Huéti
40. Ahtiago and Tobo }	Whátan	Lúin	Bánah	Kúnchi
41. Ahtiago (Alfuros) }	Ñufátanim	Lūim	Husūūm	Husum
42. Gah	Rísi	Lului	Usulah	Kuincha
43. Wahai }	Hatare	Luni	Helu	Kapai
45. Matabello	Watan	Lúru	Lóburr	Udiss
46. Teor	Telimin	Urut	Fun	Fud
49. Mysol	Badan	Kaboom	Fean	Bus
50. Mysol	Padan	Mot bom	Aan	Boo
59. Baju	Badan	Bákas	Panah	Puti

LANGUAGES OF THE MALAY ARCHIPELAGO.— *Continued.*

	BUTTERFLY.	CAT.	CHILD.	CHOPPER.	COCOA-NUT.	COLD.
1.	Kūpūkūpū	Kūching	A'nak	Párang	Klápa	Dingin,Tijok.
2.	Kūpu	Kuching	Anak	Parang	Krambil	A'dam.
6.	Kumberá	Ombutá	Oánana	Kapuru	Kalimbúngo	Magári.
7.	Kolikoti	Miaò	Anak	Berang	Nyóroh	Dingin.
16.	Karinboto	Tusa	Dodio	Kompilang	Bángoh	Madadun.
17.	Wieto	Ngeäu	Anako	Boroko	Bougo	Motimpia.
18.	KalibumbongMiau		Anak	Pedah	Bángu	Matuno.
19.		Miau	Pigi-neneh	Galéleh	Nyu.	
20.	Maápa	Nāo	Nináua	Péda	Núi	Bagóa.
21.	Lahen	Sika	A'nai	Tolie	Niwi	Numniri.
22.	Lahei	Sika	Nánat	Tódo	Niwi	Damóti.
23.	Tapalápat	Māo	ʿNaánati	Katúen	Niwi	Dabridi.
24.	Koláfi	Mau	Emlúmo	Laiey	Niwi	Komoriti.
26.	Kopa kopa	Túsa	Ngófa	Péda	Igo	Góga.
29.	Kalibobo	Tusa	Untúna	Barakas	Níwitwan	Makufin.
31.	Mimáliki	Bóki	Mangópa	Taíto	Igo	Damála.
32.	Kakópi	Túsa	Niana	Lobo	Nier	Periki.
33.	Pepeül	Sie	Wana	Lopho	Niwil	Periki.
34.	Kupo kupo	Temai	Opoliána	Ikíti	Niwéli	Mutí.
35.	Lowar lowar	Sía	Wári	Lopo	Nimil	Periki.
36.	Kokohan	Siah	Anahei	Lopo	Muŏllo	Puriki.
37.	Korūli	Maōw	Wána	Aáti	Liwéli	Pepéta.
38.		Sía	Ana	Lopo	Niwéli	Maríki.
39.	Tutupúno	Sia	Anan	Lopo	Nuélo	Pilikéko.
40.	Bubúmái	Sikar	Iniának	Béda	Núa	Bäidik.
41.		Láfim	Anavim	Tafim	Nuim	Makáriki.
42.	Kowa kowa	Shika	Dúia	Péde	Niūla	Lifie.
43.	Koháti	Sika	A'la	Tulumaina	Lúen	Mariri.
45.	Obaóba	Odára	Enéna	Béda	Dar	Arídin.
46.	Kokop	Sika	Anìk	Funén	Nōr.	Giridin.
49.	Kalabubun	Mar	Kachun	Keío	Nea.	Kabluji.
50.		Miau	Wai	Yeu	Nen	Pátoh.
59.	Titúe	Miau	Anáko	Bádi	Salóka	Jérnih.

ONE HUNDRED AND SEVENTEEN WORDS IN THIRTY-THREE

English	COME.	DAY.	DEER.	DOG.
1. Malay	Mári	A'ri (Siang.)	Rūsa	A'ujing
2. Javanese	Marein	Aivan	Rusa	Asu
6. Boutọn ⎱ S. Celebes..	Maivé	Héo	Orúsa	Muntóa
7. Salayer ⎰	Maika	Allo	Rusa	Asu
16. Menado ⎱	Simépu	Roū	Rusa	Kapuna
17. Bolang- ⎰ N. Celebes.. hitam	Aripa	Unuveno	Rusa	Ungu
18. Sanguir, Sian	Dumahi	Rókadi	Rusa	Kapúna
19. Salibabo	Maranih			Assu
20. Sula Is.	Mái	Dawíka	Munjangan	Asu
21. Cajeli ⎫	Omai	Gáwak	Mūnjángau	Aso
22. Wayapo ⎬ Bouru.	Ikomai	Dówa	Mūnjángan	Asu
23. Massaratty ⎭	Gumáhi	Liar	Munjangan	Asu
24. Amblaw	Buoma	Laei	Munjaráni	Asu
26. Tidore	Ino keré	Wellusita	Munjangan	Káso
29. Gani ⎱ Gilolo.	Mai	Balanto	Munjangan	Iyór
31. Galela ⎰	Nehíno	Taginíta	Munjangan	Gáso
32. Liang ⎫	Uimai	Kikir	Munjangan	Asu
33. Morella ⎪	Oimai	Alowata	Munjangan	Asu
34. Batumerah ⎬ Amboyna.	Omai	↘.Watiëla	Munjangan	Asu
35. Lariki ⎭	Mai	Aoaaóa	Munjangan	Asu
36. Saparua	Mai	Kai	Rusa	Asu
37. Awaiya ⎫	Alowei	Apaláwe	Maiyáni	A'su
38. Camarian ⎪	Mai		Maiyánani	Asúa
39. Teluti ⎪	Mai	Kíla	Meisakano	Wasu
40. Ahtiago and Tobo ⎬ Ceram.	Kulé	Matalima	Rúsa	Yás
41. Ahtiago (Alfuros) ⎪	Dak lápar	Pília	Tusim	Nawang
42. Gah ⎪	Mai	Malal	Rusa	Kafúni
43. Wahai ⎭	Mai	Kaseiella	Mairáran	Asu
45. Matabello	Gomári	Larnumwás	Rúsa	Afúua
46. Teor	Yef man	Liléw	Rusa	How
49. Mysol	Jog mah	Seasan	Mengangan	Yes
50. Mysol	Bo muu	Kluh	Menjangan	Yem.
59. Baju	Paituco	Lau	Paiów	Asu

LANGUAGES OF THE MALAY ARCHIPELAGO.—*Continued.*

	DOOR.	EAR.	EGG.	EYE.	FACE.	FATHER.
1.	Píntu	Telínga	Túlor	Máta	Múka	Bápa.
2.	Lawang	Kúping	U'ndok	Móto	Raì	Baba.
6.	Obámba	Talinga	Ontólo	Máta	Oroku	Amana.
7.	Pintu	Toli	Tanar	Mata	Rupa	Ama.
16.	Raroangen	Túri	Natu	Máta	Duhn	Jama.
17.	Pintu	Boronga	Natu	Mata	Paio	Kiamat.
18.	Pintu	Toli	Tuloi	Mata	Gáti	Yaman.
19.						
20.	Yamáta	Telinga	Metélo	Háma	Lúgi	Nibaba.
21.	Lilolono	Telilan	Telon	Lamūmo	Uhamo	A'mam.
22.	Káren	Telingan	Télo	Raman	Pupan	Náma.
23.	Henóloni	Linganani	Telo	Ramaui	Pupan lalin	Náama.
24.	Sowéni	Herenatia	Rehöi	Lumatibukói.	Ufnati lareni.	Amao.
26.	Móra	Ngan	Gósi	Lau	Gái	Baba.
29.	Nára	Tingēt	Toli	Umtowt	Gonaga	Bápa.
31.	Ngóra	Nangów	Magosi	Láko	Nangabío	Nambúba.
32.	Metenúre	Terina	Muntiro	Máta	Hihika	Ama.
33.	Metenulu	Telina	Mantirhui.	Mata	Uwaka	A'ma.
34.	Lamáta	Telinawa	Munteloá	Matava	Uwaro	Kopapa.
35.	Metoüru	Terina	Momatíro	Mata	U'wa	Ama.
36.	Metoro	Teréna	Tero	Mata	Wáni	Ama.
37.	Aleáni	Teríua mo	Telúli	Mata mo	Wámu mo	Ama.
38.	Metanorúi	Terinam	Terúni	Máta	Wamo	Ama.
39.	Untaniyún	Tinacóno	Tin	Matacolo	Facólo	Amacolo.
40.	Lolamatan	Likan	Tólin	Mátan	U'fan	láman.
41.	Motūlnim	Telikeinlúim.	Tolnim	Mátara	Uhúnam	Amái.
42.	Yebúteh	Tanomulino	Tolor	Matanina	Funonína	Mama.
43.	Olamatan	Teninare	Latun	Mata	Matalalin	Ama.
45.	Fidin	Tilgár	Atulú	Matáda	Omomanía	Ieí.
46.	Remátin	Karin	Telli	Matin	Matinóin	A'ma.
49.	Batal	Tenaan	Tolo	Tūn	Tunah	Mám.
50.	Bata	Mot na	Tolo	Mut morobu.	Mutino	Mām.
59.	Boláwah	Telinga	Untello	Mata	Rúa	Uáh.

I I 2

One Hundred and Seventeen Words in Thirty-three

English	FEATHER.	FINGER.	FIRE.	FISH.
1. Malay	Būlū	Jári	A'pi	Ikan
2. Javanese	Wūlu	Jári	Gúni	Iwa
6. Bouton } S. Celebes..	Owhù	Saranga	Whá	Ikáni
7. Salayer }	Bulu	Karami	Api	Jugo
16. Menado } N. Celebes..	Mombulru	Talrimido	Pūtung ‚Maranigan.	
17. Bolang-hitam }	Burato	Sagowari	Puro	Sea
18. Sanguir, Sian	Doköi	Limado	Putún	Kina
19. Salibabo			Puton	Inásah
20. Sula Is.	Nifóa	Kokowana	Api	Kéna
21. Cajeli }	Bolon	Limam kokon	Ahū	Iáni
22. Wayapo } Bouru.	Fulun	Wangan	Bána	Ikan
23. Massaratty... }	Folun	Wangan	Bána	Ikan
24. Amblaw	Buloi	Lemnati kokoli	Afu	Ikiani
26. Tidore	Gógo	Gia marága	U'ku	Nýan
29. Gani } Gilolo.	Lonko	Odeso	Lútan	Ian
31. Galela }	Ló	Rarága	Uku	Náu
32. Liang } Amboyna.	Huru	Rimaka hatu	Aow	Iyan
33. Morella }	Manuhrui	Limaka hatui	Aow	Iyan
34. Batumerah }	Hulúna	Limáwa kukualima..	Aow	Iáni
35. Lariki }	Manhúru	Lima hato	Aow	Ian
36. Saparua	Huruni	Uūn	Hao	Ian
37. Awaiya } Ceram.	Hulúe	Saäti	Aoúsa	Iáni
38. Camarian	Phulúi	Tarüni	Haɔ	Iáni
39. Teluti	Wicolo	Limacohunilo	Yáfo	Yáno
40. Ahtiago and Tobo }	Fulin	Uin	Yáf	I'an
41. Ahtiago (Alfuros)	Toholim	Tai-imara likéluni	Wáham	I'em
42. Gah	Veolūhr	Numonin tutulo	Aif	Ikan
43. Wahai }	Hulun	Kukur	Aow	Ian
45. Matabello	Alolú	Taga tagan	Efi	I'an.
46. Teor	Phulin	Limin tagin	Yaf	Ikan
49. Mysol	Guf	Kanin ko	Lap	Ein
50. Mysol	Gan	Kanin ko	Yap	Ein
59. Baju	Bolo	Eríke	Api	Déiah

Languages of the Malay Archipelago.—*Continued.*

	FLESH.	FLOWER.	FLY.	FOOT.	FOWL.	FRUIT.
1.	Dáging	Būnga	Lálah	Káki	A'yam	Būa.
2.	Dáging	Kembang	Lálah	Síkil	Pitek	Wowóan.
6.	U'ntok	Obúnga	Oráli	Oei	Mánu	Eakena.
7.	Asi	Bunga	Katinali.	Bunkin	Jangan	Bua.
16.	Gisini	Burány	Ralngoh Raédai		Mánu	Bua.
17.	Sapu	Wringonea	Rango	Teoro	Mano	Bunganea.
18.	Gusi	Lelun	Lango	Laidi	Manu	Buani..
19.					Manu	Buwah.
20.	Ni'ihi	Saía	Kafini	Yiéi	Mánu	Kao fua.
21.	Isim	Mnúrū	Bena	Bitim	Tehúi	Būan.
22.	Isin	Tatan	Féna	Kadan	Téput	Fūan.
23.	Isinini	Kao tutun	Féna	Fitinen	Téputi	Fuan.
24.	Isuatéa	Kakali	Béna	Beernyáti atani.	Rufúa	Buani.
26.	Róhe	Hatimoöto siya.	Gúphu	Yóhu	Toko	Hatimoöto sopho.
29.	Woknu	Bunga	Búbal	Wed	Manik	Sapu.
31.	Nangaláki	Mabúnga	Gúpu	Nandóhu	Tóko	Masópo.
32.	Isi	Powta	Lari	Aika	Mano	Húa.
33.	Isi	Powti	Lali	Aika	Manu	Hua.
34.	Isíva	Kahuka	Henai	Aíva.	Máno	Aihuwáua.
35.	Isi	Kupang	Pénah	Ai	Mano	Ai hua.
36.	Isini	Kupar	Upenah..	Ai	Mano hena Hwányo.	
37.	Waoúti	Lahówy	Pepénah.	Aì	Manulúma Huváiy.	
38.		Kupáni	Upéna	Ai	Mánu	Huwái.
39.	Isicolo	Tifin	Upéna	Yaicólo	Manuo	Huan.
40.	Isin	Futin	Lákar	Yái	Tófi	Vúan.
41.	Isnum	Eiheitnum	Phenem.	Wáira	Towim	Eifuanum.
42.	Sesiún	Fuis	Langar	Kaieniña	Manok	Woya.
43.	Héla	Loen	Mumun..	Ai	Malok	Huan.
45.	Ahí	Ai wói	Wéger	Owéda	Manok	Woi imotta.
46.	Henin	Pus	Omiss	Yain	Manok	Phuin.
49.	Wamut	Gáp heu		Kanin pap	Kakep	Gapeah.
50.	Mot nut	loh	Kelang.	Mat wey	Tekayap	I'po.
59.	Isi	Bunga	Langow..	Nai	Mano	Bua.

One Hundred and Seventeen Words in Thirty-three

English	GO.	GOLD.	GOOD.	HAIR.
1. Malay	Púrgi	Más	Baik	Rámbut
2. Javanese	Lungo	Mas	Butje	Rambut
6. Bouton } S. Celebes.	Lipano	Huláwa	Marápe	Bulwa
7. Salayer } S. Celebes.	Lampa	Bulain	Baji	Uhu
16. Menado } N. Celebes..	Máko	Bolraong	Sahenie	Uta
17. Bolang-hitam } N. Celebes..	Korunu	Bora	Mopia	Woöko
18. Sanguir, Sian	Dako	Mas	Mapiah, Ma holi Utan	
19. Salibabo	Ma puréteh	Bulawang	Mapyia	
20. Sula Is.	Láka	Famaká	Pía	O'ga
21. Cajeli } Bouru.	Oweho	Blawan	Ungano	Buloni
22. Wayapo } Bouru.	Iko	Balówan	Dagósa	Folo
23. Massaratty } Bouru.	Wíko	Hawan	Dagósa	Olofólo
24. Amblaw	Buoh	Bulówa	Parei	Olnáti
26. Tidore	Tagi	Guráchi	Láha	Hútu
29. Gani } Gilolo.	Tahn	Omas	Fiar	Iklet
31. Galela } Gilolo.	Notági	Gurachi	Talóha	Hútu
32. Liang } Amboyna.	Oï	Halowan	Ia	Kaiola
33. Morella } Amboyna.	Oi	Halowan	Ia	Keiúle
34. Batumerah } Amboyna.	Awái	Halowani	Amaísi	Huá
35. Lariki } Amboyna.	Oi	Halowan	Mai	Keö
36. Saparua	Ai	Halowan	Malopi	Uwóhoh
37. Awaiya } Ceram.	Aeó	Halowáni	Aólo	Uwoleíha mo.
38. Camarian } Ceram.	Aeo	Halowani	Mái	Keóri
39. Teluti } Ceram.	Itái	Hulawano	Fia	Këülo
40. Ahtiago and Tobo } Ceram.	Akó	Masa	Komúin	Ulvú
41. Ahtiago (Alfuros) } Ceram.	Teták	Masen	Komia	Ulufúim
42. Gah } Ceram.	Ketángo	Mas	Guphin	Uka
43. Wahai } Ceram.	Aou	Hulaän	Ia	Húe
45. Matabello	Fanów	Mása	Fía	U'a
46. Teor	Takek	Mas	Phien	Wultáfun
49. Mysol	Jog	Plehan	Fei	Peleah
50. Mysol	Bo	Phean	Ti	Mutlen
59. Baju	Moleh	Mas	Alla	Buli tokolo

Languages of the Malay Archipelago.—*Continued.*

HAND.	HARD.	HEAD.	HONEY.	HOT.	HOUSE.
1. Tángan........Kras.........	..Kapála........	Mádu..........	Pánas..........	Rúmah.	
2. Tángan........Kras............	U'ndass.........	Mádu............	Páuas.........	Umah.	
6. Olima.........Tobo............	Obaku.........	Ogora.........	Mopáni.......	Bánna.	
7. Lima.........Teras..........	Ulu............	Ngongonu....Bumbung	....Sapu.		
16. Rilma.........Maketihy.....Timbónang...Madu	.........Matéti	.........Balry.			
17. Rima..........Murugoso....Urie	.............Teoka	.........Mopaso	.......Bore.		
18. Lima..........Makúti	......Tumbo	..........................Matúti	........Bali.		
19.					...Bareh.
20. Lima.........Kadiga	........Nãp	.............................Baháha	......U'ma.		
21. Limámo......Namkana....Olum	.........Madu	.........Poton	.........Lúma.		
22. Fahan.........Lumé	..........Ulun fatu	.....................Dapóto	......Húma.		
23. Fahan.........Digíwi........Olun	..............................Dapótoni	......Húma.			
24. Lemnatia.....Unkiweh.....Olimbukói...Násu	.........Umpána	......Lúmah.			
26. Gia............Futúro.........Defólo...	....................Sasáhu	.........Fola.			
29. Komud......Maséti.........Poi	...................................San	............U'm.			
31. Gia.............Daputúro.....Nangasáhi...Mangópa	......Dasáho	........Táhu.			
32. Rimak.........Makána......Uruka.........Niri	............Putu	..........Rumah.			
33. Limaka......Makana......Uruka........	Keret	.........Loto............Lumah.			
34. Limáwa......Amakana....Ulúra	.............................Aputu	.........Lumá.			
35. Lima.........Makána......Uru............	Miropenah	...Pútu..........Rumah.			
36. Rimah.........Makanah.....Uru.........Madu	...........Kuno.........Rumah.				
37. A'la............Uru.........Ulu mo	......Helímah......Maoúso	......Lũũma.			
38. Limamo......Makána......Ulu	...........Násu..........Pútu	.........Luma.			
39. Limacolo......Unté..........Oyúko	..........Penanûn......Pútu	...........Uma.			
40. Niman.........Kakówan....Yúlin........Músa	..........Bafánat	......Umah.			
41. Tai-ímara.....Mocolá........Ulukátim....Lukaras	......Asála	.........Feióm.			
42. Numoniña....Kaforat......Luníni.........Nasu musun..Mofánas	......Lúme.				
43. Mimare.......Mukola......Ulure.........Kinsumi	......Mulai.........Luman.				
45. Dumada lomia Máitan..........Alúda.........Limlimur.....Ahúan.........Orúma.					
46. Limin.........Keherr.........Ulin	.............................Horip	.........Sarin.			
49. Kanin.........Umtoo........Kahutu	......Fool............Benis	.........Kom.			
50. Mot mor......Net............Mullud.......Fool............Pelah..........Dé.					
59. Tangan.......Kras............Tikolo..........................Panas	.........Rumah..				

One Hundred and Seventeen Words in Thirty-three

English	HUSBAND.	IRON.	ISLAND.	KNIFE.
1. Malay	Láki	Bŭsi	Pūlo	Písau
2. Javanese	Bedjo	Wusi	Pulo	Lading
6. Bouton ⎫ S. Celebes	Obawinena	A'sé	Liwúto	Pisau
7. Salayer ⎭	Burani	Busi	Pulo	Pisau
16. Menado ⎫	Gagijannee	Wascy	Mapuroh	Pahegy
17. Bolang- ⎬ N. Celebes. hitam ⎭	Taroraki	Oáse	Riwuto	Piso
18. Sanguir, Sian	Kapopungi	Wasi	Toadi	Pisau
19. Salibabo	Essah		Taranusa	Lari
20. Sula Is.	Túa	Mūm	Pássi	Kóbi
21. Cajeli ⎫	Umlanei	Awin	Núsa	Iliti
22. Wayapo ⎬ Bouru.	Mori	Kawil ՝.	Núsa	Irit
23. Massaratty ⎭	Gebhá	Momul	Nusa	Katánan
24. Amblaw	Emanow	Awi	Nusa	Kamarasi
26. Tidore	Nau	Búsi	Gurumongópho.	Dári
29. Gani ⎫ Gilolo.	Mondemapin	Busi	Wāf	Kobit
31. Galela ⎭	Maróka	Dodiódo.	Gurongópa	Díha
32. Liang ⎫	Mahinatima malona	Taä	Nusa	Seé
33. Morella ⎬ ₐₘᵦₒᵧₙₐ. Amolono	Amolono	Ta	Nusa	Seéti
34. Batumerah ⎬ Mundai	Mundai	Saëi	Nusa	Opiso
35. Lariki ⎭ Malona	Malona	Mamōr	Nusa	Séi
36. Saparua	Manowa	Mamōlo	Nusa	Seit
37. Awaiya ⎫	Manowai	Mamóle	Mísa	Amasáli
38. Camarian	Malóna	Mamóle	Nusa	Seíti
39. Teluti	Ihina manowa	Momollo	Nusa	Seito
40. Ahtiago and Tobo ⎬ Ceram.	Imyóna	Momūm	Túbil	Tuána
41. Ahtiago (Alfuros)	Ifnéinin sawanim	Momolin	Tuplim	Macouosim.
42. Gah	Bulana	Momúmi	Tubur	Tuka
43. Wahai ⎭	Pulahan	Héta	Lusan	Tuluangan.
45. Matabello	Helameranna	MomúmoTobūr		Mirass
46. Teor	Wehoin	Momúm	Lowánik	Isowa
49. Mysol	Man	Seti	Yef	Cheni
50. Mysol	Mot man	Leti	Ef	Yeaói
59. Baju	Ndáko	Bisi	Pulow	Pisau

Languages of the Malay Archipelago.—*Continued.*

	LARGE.	LEAF.	LITTLE.	LOUSE.	MAN.	MAT.
1.	Būsar ......Daūn...		.............Kíchil	........Kūtū .....'..	Orang lákilakiTíkar.	
2.	Gedé....... Godong......		......Chilí	.........Kūtu	Wong lanan.. Klosso.	
6.	Moughí ...Tawána	.............Kidikidi	......Okútu	......Omani.... ....Kiwaru.		
7.	Bákeh .....Taha	.............Kédí	.........Kutu	......Tau	.........Tupur.	
16.	Raboh .....Daun	.........Dodio	.........Kutu	......Taumata esenSapie.		
17.	Morokaro..Lungianea ·.........Moisiko	......Kutu	......Roraki	........Boraru.		
18.	Labo........Decaluni	.............Aníou	.........Kutú	..... Manesh......	..Sapieh.	
19.	Bagewa ....Kadodo.......................Tomatá	...... Bilátah.				
20.	Eá...........Kao hósa...........Mahé	.........Kóta	......Maona.........Saváta.			
21.	Léhai ......Atétun.................Koi	...........Olta	........Umlanai	......A'pine.		
22.	Bágut ......Kroman	.... Roit	............Kóto	......Gemana	.......A'tin.	
23.	Haat........Kúman.................Roi	............Koto	......Anamhána ...Kátini.			
24.	Plaré ......Lai obawai	........Bakoti	.........Uru..........Remau.........Arimi.			
26.	Lámu .....Hatimooto merow. Kéni	..........Túma	......Nonán	........Junnito.		
29.	Talalólo ....Nilonko	............Waiwáio	.....Kútu	......Mon	...........Kalása.	
31.	Elámo .....Misóka ....Dechéki	......Gáni.........Anów	.........Jungúto.			
32.	Nila........Ailow	............Koi	...........Utu	......Malona........Pai.		
33.	Hella......Aịlow	.............Ahúntai	......Utu	.........Malono	.......Hilil.	
34.	Enda-a.....Aitéti	............Aná-á	.........Utu	......Mundai	.......Towai.	
35.	Ira ....Ai rawi	.............Koi	..........Kutu	......Malona	......Paíl.	
36.	Ilahil ......Laun	............Ihíhil	.........Utu	......Tumata	.......Pai.	
37.	Iláhe ......Laíni	..............Olíhil..........U'tu ... .Tumata	......Kaili.			
38.	Eráámei ...Airówi..............Kokaneii......Utúa........Tumata	......Paílí.				
39.	Elau........Daun ....Anan	.........Utu	......Manusia	......Pai-ilo.		
40.	Aíyuk .....Lan	.................Nélak	.........Tínan	......Muána	........Láb.	
41.	Poten.......Eilúnim ....'.Anaanịn	......Kutim	.....Muruleinum..Lapim.			
42.	Bobuk .....Liuo.................Wota wota ...Kutu	......Beláne.........Kiël.				
43.	Máina .....Totun	..............Kiiti............Utun	......Ala híciti ....Kihu.			
45.	Leléh ......Arehín.............Enena	.........U'tu	........Marananna ...I'ra.			
46.	Lēn ....Chafen.............Fek	...........Hut	.........Meránna	......Fira.	
49.	Sala ....Kaluin.............Gunam	........Ut	.........Motu	.........Tin.	
50.	Klen ....Idun ....Senpoh	.......Uti	.........Mot	...........Tin.	
59.	Basar ......Daun ....Didıki	.........Kutu	......Lélah	.........Tepoh.	

ONE HUNDRED AND SEVENTEEN WORDS IN THIRTY-THREE

English		MONKEY.	MOON.	MOSQUITO.	MOTHER.
1. Malay		Mūnyeet	Būlan	Nyámok	Ma
2. Javanese		Budéss	Wulan	Nyámok	Mbo
6. Bouton	S. Celebes	Róke	Búla	Burótok	Inaná
7. Salayer	S. Celebes	Dáre	Bulan	Kasisili	Undo
16. Menado	N. Celebes	Bohen	Bulrang	Tenie	Inany
17. Bolang-hitam	N. Celebes	Kurango	Wura	Kongito	Leyto
18. Sanguir, Sian		Babah	Buran	Túni	Inúngi
19. Salibabo			Burang		
20. Sula Is.		Mía	Fasina	Samábu	Nieía
21. Cajeli	Bouru.	Kessi	Būlani	Suti	Inámo
22. Wayapo	Bouru.	Kess	Fhūlan	Múmun	Neína
23. Massaratty	Bouru.		Fhulan	Seúgeti	Neína
24. Amblaw		Kess	Bular	Sphúre	Ina
26. Tidore		Mía	O'ra	Sisi	Yaíya
29. Gani	Gilolo.	Nok	Pai	Nini	Mamo
31. Galela	Gilolo.	Mía	O'sa	Gumóma	Maówa
32. Liang	Amboyna.	Sia	Hulanita	Séne	Ina
33. Morella	Amboyna.	Aruka	Hoolan	Sisil	Inaö
34. Batumerah	Amboyna.	Késs	Huláni	Sisili	Inao
35. Lariki	Amboyna.	Rúa	Haran	Sūn	Ina
36. Saparua		Rua	Phulan	Sonot	Ina
37. Awaiya	Ceram.	Kesi	Phuláni	Manisíe	Ina
38. Camarian	Ceram.	Kesi	Wuláni	Senóto	Ina
39. Teluti	Ceram.	Lúka	Hiáno	Sumóto	Inaú
40. Ahtiago and Tobo	Ceram.	Lūkar	Phúlan	Minís	Aína
41. Ahtiago (Alfuros)	Ceram.	Meiram	Melim	Manis	Inái
42. Gah	Ceram.	Lēk	Wúan	Umiss	Nina
43. Wahai	Ceram.	Yakiss	Hulan	U'muti	Ina
45. Matabello		Léhi	Wúlan	U'muss	Nína
46. Teor		Lok	Phulan	Rophun	I'na
49. Mysol			Pet	Kamumus	Nin
50. Mysol			Náh	Owei	Nin
59. Baju		Mondo	Bulan	Sisil	Máko

Languages of the Malay Archipelago.—*Continued.*

	MOUTH	NAIL (FINGER)	NIGHT	NOSE	OIL	PIG
1.	Múlūt	Kúkū	Málam	Idong	Mínyak	Bábi.
2.	Sánkum	Kūku	Bungi	Irong	Lūngo	Chilong.
6.	Nánga	Kuku	Maromó	Oánu	Mínak	Abáwhu.
7.	Bawa	Kanuko	Bungi	Kumor	Minyak	Bahi.
16.	Mohong	Kanuku	Máhri	Hidong	Rana	Babi.
17.	Nganga	Kamiku	Gubie	Jjunga	Rana	Rioko.
18.	Mohon	Kanuko	Hubbi	Hirong	Lana	Bawi.
19.						Bawi.
20.	Beióni	Kowóri	Bohúwi	Né	Wági	Fafi.
21.	Nūūm	Uloimo	Petū	Nem	Nielwíne	Babúe.
22.	Muen	Utlobin	Béto	Nien	Newiyn	Fafu.
23.	Naónen	Logini	Béto	Nioni	Newiny	Fafú.
24.	Numátéa	Heruenyati	Pirue	Neínya téha.	Nivehoi	Bawu.
26.	Móda	Gulichiti	Sophúto	Ûn	Guróho	Sóho.
29.	Sumut	Kuyut	Becómo	Usnut	Nimósu	Boh.
31.	Nangúru	Gitipi	Daputo	Ngúno	Gosóso	Titi.
32.	Hihıka	Terèina	Hatóru	Hırúka	Neerwiyn	Hahow.
33.	Soöka	Tereiti	Hatolu	lúka	Neerliyn	Hahu.
34.	Suara	Kuku	Hulamti	Ninúra	Wakéli	Hahu.
35.	Ihi	Terein	Halometi	I'ru	Nimimein	Hahu.
36.	Nuku	Teri	Potu	Iri	Warisini	Hahul.
37.	Ihi mo	Talü	Mute	Nua mo	Wailasmi	Hāhu.
38.	So		Améti	Hilimo	Wailisini	Hawhúa
39.	Hihico	Talicólo	Humoloi	Olicolo	Fofótu	Hahu.
40.	Vudin	Selíki	Matabūt	I'hn	Kūl	Wār.
41.	Tafurnum		Potūūn	I'huum	Félim	Fafuim.
42.	Lonina	Wuku	Garagaran	Sonına	Gúa	Bóia.
43.	Siurure	Talahikun	Manemi	Iuore	Héli	Hahu.
45.	Ilida	Asitiggir	Olawáha	Werámani	Gúla	Boör.
46.	Huin	Limin kukin	Pogaragara	Gilinkani	Hīp	Faf.
49.	Gulan	Kasebo	Maléh	Shong gulu	Majulu	Boh.
50.	Mot po	Kok nesib	Mau	Mot mobi	Menik	Boh.
59.	Boah	Kuku	Sangan	Uroh	Mángo	Góh.

ONE HUNDRED AND SEVENTEEN WORDS IN THIRTY-THREE

English	POST.	PRAWN.	RAIN.	RAT.
1. Malay	Tíeng	Udong	Hūjan	Tíkus
2. Javanese	Soko	Uran	Hudan	Tikus
6. Bouton } S. Celebes	Otúko	Meláma	Waó	Bokóti
7. Salayer }	Palayaran	Doün	Bosi	Blaha
16. Menado } 17. Bolang- } N. Celebes hitam }	Dihi	Udong	Tahíty	Barano
	Panterno	Ujango	Oha	Borabu
18. Sanguir, Sian	Dihi	Udong	Tahiti	Balango
19. Salibabo	Pari-arang		Urong	
20. Sula Is.	H'ii	U'ha	Húya	Saáfa
21. Cajeli }	Ateoni	Ulai	U'lani	Boti
22. Wayapo } Bouru	Katehan	Uran	Dekat	Boti
23. Massaratty }	Katéheni	Uran	Dekati	Tíkuti
24. Amblaw	Hampowne	Ulai	Ulah	Púe
26. Tidore	Ngasu	Búrowi	Béssar	Múti
29. Gani } Gilolo 31. Galela }	Li	Níke	Ulan	Lūf
	Golingáso	Dódi	Húra	Lúpu
32. Liang	Riri	Méter	Hulan	Maláha
33. Morella	Lili	Metar	Hulan	Malaha
34. Batumerah	Lili	Metáli	Huláni	Puéni
35. Lariki	Leilein	Mítal	Haran	Maláha
36. Saparua	Riri	Mital	Tiah	Mulahah
37. Awaiya	Lili	Mitáli	Uláne	Maláha
38. Camarian	Lili	Mitali	Uláni	Maláha
39. Teluti	Hili	Mutáyo	Gia	Maiyáha
40. Ahtiago and Tobo	Fólan	Filúan	U'lan	Meláva
41. Ahtiago (Alfuros)	Faolnim	Hoim	Roim	Sikim
42. Gah	Usa	Gurun	U'an	Karúfei
43. Wahai	Hinin	Bokoti	Ulan	Mulahan
45. Matabello	Faléra	Gúrun	Udáma	Arófa
46. Teor	Pelérr	Gurun	Hurani	Fudarúa
49. Mysol	Fohan	Kasána	Golim	Keluf
50. Mysol	Felian	Kasana	Golim	Quóh
59. Baju	Tikala	Dóah	Huran	Tikus

Languages of the Malay Archipelago.— *Continued.*

	RED.	RICE.	RIVER.	ROAD.	ROOT.	SALIVA.
1.	Mérah ....Brās........Sūngei........Jálan ....A'kar..........Lúdah.
2.	Abang ....Bras........Sungei........Malaku ......Oyok ....I'du.
6.	Meräi ....Bai........Uvé ....Dára ....Koleséna ......Ovilu.
7.	Eja.............Biras ......Balang........Lalan........Akar ....Pedro.
16.	Mahamu ......Bogáseh ...Raríou........Dalren ....Hámu.........Edu.
17.	Mopoha.......Bugasa.....Ongagu ......Lora...........Wakatia ......Due.
18.	Hamu ....Bowáseh ...Sawán ....Dalin........Pungenni.....Udu.
19.	Maramutah...Boras.
20.	Mia ....Bíra ....Sungei........A'ya...........Kao akar......Bihú.
21.	Unmíla ......Hálai........Wai lé........Lalani........Alamúti ......Bulai.
22.	Míha ....Hála ....Wai fatan ...Tuhun.... ....Púhah.
23.	Miha ....Pála........Wai ....Tóhoni.........Kao lahin ....Fúhah.
24.	Meháni ......Fála........Waibatang ...Lahuléa......Owáti... ......Rubunatéa.
26.	Kohóri........Bira........Wai ....Lolinga ......Hatimoöto ...Gidi.
29.	Mecoit ......Samasi......Waiyr ....Lolan ....Niwolo...... ...Iput.
31.	Desoélla .....Itámo .....Siléra.........Néko ....Kiví.
32.	Kao ....Allar ......Weyr ....Lahan ....Waäta ....Tehula.
33.	Kao ....Allar ......Weyl hatei...Lalan ....Eiwaäti........Tehula.
34.	Awow ....Allái ....Laláni ....Ai ....Tohulá.
35.	Kao ....Hála ....Wai hatei ...Lalan ....Ai waat......Tohural.
36.	Kao ....Hálal ......Walil.........Lalano ....Aiwaári......Tohulah.
37.	Meranáte......Hála ....Waliláhe ....Laláni ....Lamúti ......Tohulah.
38.	Kaō ....Hála ....Waliráhi .....Lalani ....Haiwaári .....Tohúlah.
39.	Kao ....Fála ....Wailolún......Latína ....Yai ....Apícolo.
40.	Dadow ....Fála ....Wailálan......Lólan ....(Ai) waht.....Béber.
41.	Lahanín ......Hálim ......Wailanim....Lalim ....Ai liléham....Píto.
42.	Merah ....Faasi ....Arr lehn .....Lāān ....Akar ....Gunísia.
43.	Mosina........Allan........Tolo maina ...Olamatan ....Tamun........Aito.
45.	Ulúli ..'.......Fáha........Arr sūasūa ...Laran ....Ai áha........Ananihi.
46.	Fulifúli ......Paser ...... Wehr fofowt .Lagain ....Woki ....Munini.
49.	Mamé ....Fās ....Wayr..........Lelin ....Gaka watu ...Clif.
50.	Shei ....Fās .... Weyoh.........Má ....Aikówa ......Tefoo.
59.	Merah ....Buas ....Ngusor ......Lalan .... Lijah.

One Hundred and Seventeen Words in Thirty-three

English		SALT.	SEA.	SILVER.	SKIN.
1. Malay		Gáram	Laut	Pérak	Kúlit
2. Javanese		Uyah	Segóro	Perak	Kūlit
6. Bouton	⎱ S. Celebes.	Gára	Andal	Riáli	Okulit
7. Salayer	⎰	Sela	Laut	Salaka	Balulan
16. Menado	⎰ N. Celebes.	Asing	Sási	Salraka	Pisy
17. Bolang-hitam	⎰	Simuto	Borango	Ringit	Kurito
18. Sanguir, Sian		Asing	Laudi	Perak	Pisi
19. Salibabo			Tagaroang	Salaba	Timokah
20. Sula Is.		Gási	Mahi	Salaka	Koli
21. Cajeli	⎱	Sasi	Olat	Siláka	Usum
22. Wayapo	⎰ Bouru.	Sasi	Olat	Siláka	Usam
23. Massaratty	⎰	Sasi	Masi	Silaka	Okonen
24. Amblaw		Sasieh	Lanti	Silaka	Tinyau
26. Tidore		Gási	Nólo	Saláka	A'hi
29. Gani	⎱ Gilolo.	Gási	Wólat	Salaka	Kakutut
31. Galela	⎰	Gási	Teow	Salaka	Makáhi
32. Liang	⎱	Tasi	Mit	Pisiputi	Urita
33. Morella	⎰ Amboyna.	Tasi	Met	Salaka	Uliti
34. Batumerah	⎰	Tási	Lauti	Salaka	Asáva
35. Lariki	⎰	Tasi	Lautan	Salaka	U'sa
36. Saparua		Tasi	Sawah	Salaka	Kutai
47. Awaiya	⎱	Tasíe	Lauhaha	Saláka	Lelutini
48. Camarian	⎰	Tasíe	Lauhaha	Salaka	Wehúi
49. Teluti	⎰ Ceram.	Lósa	Toweín	Salák	Lilicolo
30. Ahtiago and Tobo	⎰	Másin	Tási	Salaka	Ikulit
31. Ahtiago (Alfuros)	⎰	Teísim	Taisin	Salaka	
32. Gah	⎰	Síle	Tasok	Salak	Likito
43. Wahai	⎰	Tasi	Laut	Seláka	Unin
45. Matabello		Síra	Táhi	Saláha	Aliti
46. Teor		Siren	Hoak	Silaka	Holit
49. Mysol		Lesin	Sol	Sulūp	Kine
50. Mysol		Garam	Belot	Salup	Mot kehin
59. Baju		Garam	Medilaut	Salaka	Kulit

LANGUAGES OF THE MALAY ARCHIPELAGO.—*Continued.*

	SMOKE.	SNAKE.	SOFT.	SOUR.	SPEAR.	STAR.
1.	A'sap	Ū'lar	Lúmbūt	Másam	Tómbak	Bíntang.
2.	Kukos	Ulo	Gárno	A'sam	Tombak	Lintang.
6.	Ombu	Sávha	Marobá	Amopára	Pandáno	Kalipopo.
7.	Minta	Saa	Lumut	Kusi	Poki	Bintang.
16.	Pūpūsy	Katoün	Marobo	Maresing	Budiak	Bitūy.
17.	Obora	Noso	MurumpitoMorosomo			Matitie.
18.		Katóan	Musikomi	Naloso	Malehan	Bitúin.
19.						Kanumpitah.
20.	Apfé	Túi	Maóma	Maníli	Pedwihi	Fatúi.
21.	Melūn	Nehei	Namlomo	Numnino	Tombak	Tūlin.
22.	Fénen	Níha	Lómo	Dumílo	Néro	Tūlu.
23.	Fenen	Wao	Lumlóba	Dumwilo	Nero	Tólóti.
24.	Mipéli	Nife	Maloh	Numliloh	Tuwáki	Maralai.
26.	Munyépho	Yéya	Bóleh	Logi	Sagu sagu	Ngóma.
29.	Iáso	Bow	Iklūt	Manil	Sagu-sagu	Betól.
31.	Odópo	Inhíar	Damúdo	Dakíopi	Tombak	Ngóma.
32.	Kunu	Nia	Apoka	Marino	Taha	Marin.
33.	Aowaht	Nia	Polo	Marino	Túpa	Marin.
34.	Asaha	Niéi	Maluta	Amokinino	Sapolo	Alanmatána.
35.	Aow pōt	Niar	Máro	Marino	Topar	Mari.
36.	Poho	Niar	Maru	Marimo	Kalēi	Mareh.
37.	Weíli	Tepéli	Mamoúni	Maalino	Soláni	Oōna.
38.	Poòti	Nía	Máru	Maaríno	Sanóko	Umáli.
39.	Yafoin	Nifar	Málu	Malim	Tupa	Meléno.
40.	Numi	Búfin	Mamálin	Manil	Túba	Tói.
41.	Waham rapoi	Koioim	Mulisním	Kounim	Leis-ánum	Kohim.
42.	Kobun	Tekoss	Malúis	Mateìbi	Oika	Tilassa.
43.	Honin	Tipolum	Mulumu	Manino	Tite	Teën.
45.	Ef ubun	Tofágin	Malúis	Matílū	Galla galla	Tóin.
46.	Yaf mein	Urubai	Máfon	Metiloi	Gala gala	Tokun.
49.	Las	Pok	Umblo	Embisin	Chei	Toen.
50.	Yap hoi	Pok	Rum	Pep	Dei	Náh.
59.	Umbo	Ular	Lúmah	Gúsuh	Wijah	Kúliginta.

One Hundred and Seventeen Words in Thirty-three

	English		SUN.	SWEET.	TONGUE.	TOOTH.
1.	Malay		Máta-ári	Mánis	Lídah	Gígi
2.	Javanese		Sungingi	Lūgi	I'lat	U'ntu
6.	Bouton	S. Celebes.	Soremo	Maméko	Lilah	Nichi
7.	Salayer	S. Celebes.	Mata-alo	Tuni	Lilah	Gigi
16.	Menado	N. Celebes.	Mata roú	Manisy	Lilah	Ngisi
17.	Bolang-hitam	N. Celebes.	Unu	Mogingo	Dila	Dongito
18.	Sanguir, Sian		Kaliha	Mawangi	Lilah	Isi
19.	Salibabo		Allo			
20.	Sula Is.		Léa	Mína	Máki	Níhi
21.	Cajeli	Bouru.	Léhei	Enmínei	Mahmo	Nisim
22.	Wayapo	Bouru.	Hangat	Dumína	Maan	Nisi
23.	Massaratty	Bouru.	Lia	Durianaa	Maanen	Nisinen
24.	Amblaw		Laei	Mina	Munartéa	Nisnyatéa
26.	Tidore		Wángi	Mámi	Aki	Ing
29.	Gani	Gilolo.	Fowé	Gamis	Imōd	Afod
31.	Galela	Gilolo.	Wangi	Damúti	Nangaládi	Ini
32.	Liang	Amboyna.	Riamata	Masusu	Meka	Nikï
33.	Morella	Amboyna.	Liamátei	Masusu	Méka	Nikin
34.	Batumerah	Amboyna.	Limatáni	Kaséli	Numáwa	Nindiwa
35.	Lariki	Amboyna.	Liamáta	Masúma	Méh	Niki
36.	Saparua		Riamatani	Mosuma	Me	Nio
37.	Awaiya	Ceram.	Líamateí	Emási	Méi	Nisi mo
38.	Camarian	Ceram.	Liamatei	Masóma	Meëm	Nikim
39.	Teluti	Ceram.	Liamatan	Sunsúma	Mecólo	Lilico
40.	Ahtiago and Tobo	Ceram.	Liamátan	Merasan	Mélin	Nifan
41.	Ahtiago (Alfuros)	Ceram.	Léum		Nínum	Nesnim
42.	Gah	Ceram.	Woleh	Masárat	Lemukonina	Nisikonina
43.	Wahai	Ceram.	Leān	Moleli	Me	Lesin
45.	Matabello		Olēr	Mateltelátan	Tumomá	Nifóa
46.	Teor		Lew	Minek	Mēn	Nifin
49.	Mysol		Seasan	Krismis	Aran	Kalifin
50.	Mysol		Kluh	Mis	Aran	Kelif
59.	Baju		Matalon	Manis	Délah	Gigi

Languages of the Malay Archipelago.—*Continued.*

	WATER.	WAX.	WHITE.	WIFE.	WING.	WOMAN.
1.	A'yer........Lílin	........Pūtih	..........Bíni	............Sayap....	.....Purumpuan.
2.	Banyu......Lílin	......Puté............	Seng wedo	...Sewíwi........	Wong wedo.
6.	Mánu......Taru	........Mapúti	.......Orakenana	...Opáni	........Bawíne.
7.	Aer ....Pantis	......Putih	........Baini....:....	Kapi...	........Baini.
16.	Akéi........Tadu	......Mabida	......Gagijan	......Panidey	......Taumatababiney.
17.	Sarúgo ....Tajo	........Mopotiho	.....Wure	........Poripikia	......Bibo.
18.	Aki ....Lilin	......Mawirah	.....Sawa	.........Tula	..........Mahoweni.
19.	Wai......................	Mawirah	......Babineh	......................Babineh.
20.	Wai........Tócha	......Boti	............Nifáta	........Sóba..........	.Fina.
21.	Wäili......Lilin	......Umpóti	......Sówom........	Ahiti	.Umbinei.
22.	Wai......................	Bóti	............Gefína	........Ahit	.....Gefíneh.
23.	Wai......................	Bóti	............Fínha	........Panìn	........Fíneh.
24.	Wai........Lilin	......Purini	........Elwinyo	......Aféti	.Remau elwınyo.
26.	Aki ....Tóeha	......Bubúlo	........Foyá	..........Fila fila	......Fofoyá.
29.	Waiyr......Tócha	......Wulan	........Mapìn	........Nifako	........Mapìn.
31.	Aki ....Tócha	......Daári	..........Mapidéka	....Gulupúpo	....Opedéka.
32.	Weyr......Kina	......Putih	........Mahina	......Aïna...........	.Mahina.
33.	Weyl ......Lilin	........Putih	........Mahina	...Ihóti	.Mahina.
34.	Weyl	Putih	...Mahinai	......Kihoá	...Mainai.
35.	Weyl ......Lilin	........Putih	........Mahina	......I'ho	.Mahina.
36.	Wai........Riruiah	....Putil	..........Pipina	........Ihol	.Pipina.
37.	Wäéli ......Lilin ...	......Putíle	........Mumahéna	...Teyhóli ...	...Mahína.
38.	Wäéli ......Lilin ...	......Putih	........Nímahína	....Ihóri	.Mahina.
39.	Wélo ......Nínio	......Putih	........Nihina........	.Hihóno	.Ihina.
40.	Waí........Lilin ...	......Babút	........Invína	......Yeón	.Vína.
41.	Wai-im	Putih	........Ifnéinin	................Ifnéinin.
42.	Arr ....Lilin	......Maphutu ..	..Bina ....Wákul	.......Binëi.
43.	Tólun ......Lilin	........Puteh	..........Pinan	..........Keheil	........Pina híeti.
45.	Arr ....Lilin	......Maphúti	......Ahéhwá	......Olilífi	........Felelára.
46.	Wehr......................	Sélup	..........Wewina	......Fanik	......Mewina.
49.	Wayr ......Telilin	......Bus	..........Pin	Kufeu	........Pin.
50.	....	Boo	............Ji yu	........Fieh ....Mot yu.
59.	Boi...................	.....Potih	..........Lako	.........Kapéna	.......Dindah.

One Hundred and Seventeen Words in Thirty-three

English	WOOD.	YELLOW.	ONE.	TWO.
1. Malay	Káyū	Kūning	Sátu	Dúa
2. Javanese	Kayu	Kuning	Sa, Sawiji	Loro
6. Bouton } S. Celebes.	Okao	Mákuni	Saangu	Ruano
7. Salayer }	Kaju	Didi	Sedri	Rua
16. Menado } 17. Bolang- } N. Celebes.	Kalun	Madidihey	Esa	Dudua
hitam }	Kayu	Morohago	Soboto	Dia
18. Sanguir, Sian	Kalu	Ridihi	Kusa	Dua
19. Salibabo	Kalu	Maririkah	Sembäow	Dua
20. Sula Is.	Kaō	Kuning	Hía	Gahú
21. Cajeli }	Aow	Umpóro	Silei	Lua
22. Wayapo } Bouru.	Kaō	Konin	Umsiun	Rua
23. Massaratty... }	Kaō	Koni	Nosiúni	Rua
24. Amblaw	Ow	Umpotoi	Sabi	Lua
26. Tidore	Lúto	Kuráchi	Remoi	Malófo
29. Gani } Gilolo.	Gagi	Madímal	Lepso	Leplu
31. Galela }	Góta	Decokuráti	Moi	Sinuto
32. Liang } Amboyna.	Ayer	Poko	Sa	Rua
33. Morella }	Ai	Poko	Sa	Lua
34. Batumerah }	Ai	Apoo	Wása	Luá
35. Lariki }	Ai	Poko	Isa	Dua
36. Saparua	Ai	Pocu	Esa	Rua
37. Awaiya }	Ai	Poporóle	Lai-isa	Lūūa
38. Camarian	Ai	Pocu	Isái	Lúa
39. Teluti } Ceram.	Lyeií	Poko	San	Lua
40. Ahtiago and Tobo }	A'i	Ununing	San	Lua
41. Ahtiago (Alfuros)	Ai-im	Uninim	Esá	Elúa
42. Gah	Kaya	Kunukunu	So	Lotu
43. Wahai }	Ai	Masikuni	Sali	Lua
45. Matabello	A i	Wuliwulan	Sa	Rua
46. Teor	Kai	Kúni	Kayée	Rúa
49. Mysol	Gáh	Kumenis	Katim	Lu
50. Mysol	Ei	Flo	K'tim	Lu
59. Baju	Kayu	Kuning	Sa	Dua

Languages of the Malay Archipelago.—*Continued.*

	THREE.	FOUR.	FIVE.	SIX.	SEVEN.	EIGHT.
1.	Tíga	A'mpat	Líma	A'nam	Tújoh	Delápau
2.	Talu	Papat	Lima	Nanam	Pitu	Wolu.
6.	Taruáno	Patánu	Limánu	Namano	Pituáno	Veluáno.
7.	Tello	Ampat	Lima	Unam	Tujoh	Karna.
16.	Tateru	Pa	Rima	Num	Pitu	Walru.
17.	Toro	Opato	Rima	Onomo	Pitu	Waro.
18.	Tellon	Kopa	Lima	Kanum	Kapitu	Walu.
19.	Tetálu	Apátaħ	Delima	Annuh	Pitu	Waru.
20.	Gatíl	Gariha	Lima	Gané	Gapítu	Gatahúa.
21.	Tello	Há	Lima	Ne	Hito	Walo.
22.	Tello	Pá	Lima	Né	Pito	Etrúa.
23.	Tello	Pa	Lima	Né	Pito	Trúa.
24.	Relu	Faä	Lima	Noh	Pitu	Walu.
26.	Rangi	Ráha	Runtóha	Rora	Tumodí	Tufkángi.
29.	Leptol	Lepfoht	Leplim	Lepwonan	Lepfit	Lepwal.
31.	Sängi	Iha	Matóha	Butánga	Tumidingi	Itupangi.
32.	Tero	Hani	Rima	Nena	Itu	Waru.
33.	Telo	Hata	Lima	Nena	Itu	Waru.
34.	Telua	Atá	Limá	Nená	Ituá	Walúa.
35.	Toro	Aha	Rima	Nóo	Itu	Waru.
36.	Toru	Haä	Rima	Noöh	Hitu	Waru.
37.	Te-elu	Aäta	Lima	Nōme	Witu	Walu.
38.	Tello	A'ä	Lima	Nome	Itu	Walu.
39.	Toi	Fai	Lima	Noi	Fitu	Wagu.
40.	Tōl	Fet	Lima	Num	Fīt	Wal.
41.	E'ntol	Enháta	Enlima	Ennói	Enhit	Enwol.
42.	Tolo	Faat	Lim	Wonen	Fiti	Alu.
43.	Tolo	Ati	Nima	Lomi	Itu	Alu.
45.	Tolu	Fata	Rima	Onaɯn	Fitu	Allu.
46.	Tel	Faht	Lima	Nem	Fit	Wal.
49.	Tol	Fut	Lim	Onum	Fit	Wal.
50.	Tol	Fut	Lim	Onum	Tit	Wal.
59.	Tiga	Ampat	Lima	Nam	Tujoh	Dolapan.

One Hundred and Seventeen Words in Thirty-three

	English		NINE.	TEN.	ELEVEN.
1.	Malay		Sambílan	Sapúloh	Sapúloh sátu
2.	Javanese		Sanga	Pulah	Swalas
6.	Bouton	S. Celebes.	Sioánu	Sapúloh	Sapúloh sano
7.	Salayer	S. Celebes.	Kasa	Sapuloh	Sapuloh sedrú
16.	Menado	N. Celebes.	Sio	Mapulroh	
17.	Bolang-hitam	N. Celebes.	Sio	Mopuru	
18.	Sanguir, Sian		Kasiow	Kapuroh	Mapurosa
19.	Salibabo		Sioh	Mapuroh	Ressa
20.	Sula Is.		Gatasía	Póha	Poha di hia
21.	Cajeli	Bouru.	Siwa	Boto	Boto lesile
22.	Wayapo	Bouru.	Eshía	Polo	Polo geren en sium
23.	Massaratty	Bouru.	Chía	Polo	Polo tem sia
24.	Amblaw		Siwa	Buro	Buro lani sebi
26.	Tidore		Sio	Nigimói	Nigimói seremoi
29.	Gani	Gilolo.	Lepsiu	Yagimso	Yagimso lepso
31.	Galela	Gilolo.	Sio	Megió	Megió demoi
32.	Liang	Amboyna.	Sia	Husa	Huséla
33.	Morella	Amboyna.	Siwa	Husá	Huselali
34.	Batumerah	Amboyna.	Siwá	Husa	Husalaisa
35.	Lariki	Amboyna.	Siwa	Husa	Husaelel
36.	Saparua		Siwa	Husani	Husani lani
37.	Awaiya	Ceram.	Siwa	Hutūsa	Sinleūsa
38.	Camarian	Ceram.	Siwa	Tineín	Salaise
39.	Teluti	Ceram.	Siwa	Hútu	Mesileë
40.	Ahtiago and Tobo	Ceram.	Siwa	Vūta	Vut säilan
41.	Ahtiago (Alfuros)	Ceram.	Ensiwa	Fotusa	Fotusa elése
42.	Gah	Ceram.	Sia	Ocha	Ocha le se
43.	Wahai	Ceram.	Sia	Husa	Husa lesa
45.	Matabello		Sia	Sow	Terwahei
46.	Teor		Siwer	Hutá	Ocha kilu
49.	Mysol		Si	Lafu	Lafu kutim
50.	Mysol		Sin	Yah	Yah tem metim
59.	Baju		Sambilan	Sapuloh	

LANGUAGES OF THE MALAY ARCHIPELAGO. — *Continued.*

	TWELVE.	TWENTY.	THIRTY.	ONE HUNDRED.
1.	Sapúloh dúa.............	Dúa pūloh	Tiga pūloh.........	Sarátus.
2.	Rolas	Rongpuluh	Talupuluh	Atus.
6.	Sapúlohruano...........	Ruapulo	Tellopulo...........	Sáatu.
7.	Sapuloh rua.............	Ruampuloh........	Tellumpuloh	Sabilangan.
16.	....			...Mahasu.
17.	....			Gosoto.
18.	Mapuro dua.............	Duampuloh	Tellumpulo.........	Mahásu.
19.	Ressa dua...............	Dua puroh	Tetalu puroh.......	Ma rasu.
20.	Poha di gahú............	Poha gahú	Poha gatíl.........	O'ta.
21.	Betele dua	Botlua...	Bot telo............	Bot ha.
22.	Polo geren rua...........	Porúa	Potéllo.........	U'tun.
23.	Polo tem rua.............	Porúa	Potello.............	U'tun.
24.	Bōr lan lua	Borolua.............	Borélo	Uruni.
26.	Nigimói semolopho ....Neginnelopho.......		Negerangi	Ratumoi.
29.	Yagimsoleplu............	Yofalu..............	Yofatol	Utinso.
31.	Megió desinoto	Menohallo....	Muruangi	Rátumoi.
32.	Husa lua................	Huturúa	Hutáro	Hutúna.
33.	Husa lua................	Huturua	Hutatilo	Hutūn.
34.	Husalaisa lua...........	Hotulua	Hotelo..............	Hutunsá.
35.	Husendua	Hutorua	Hutóro	Hutūn.
36.	Husani elarua...........	Huturua	Hutoro	Utúni.
37.	Sinlūa	Hutulúa	Hututēlo	Utúni.
38.	Salalua..................	Hutulua	Hututello	Hutunére.
39.	Hutulelúa	Hutulúa ....Hututoi		Hutún.
40.	Vut sailan lūa...........	Vut lua..	Vut tol	Utin.
41.	Elelúa	Fotulúa	Fotol	Hutnisá.
42.	Husa la lua..............	Otoru.......	Otólu	Lutcho.
43.	Ocha siloti	Hutu a......	Hutu tololu	Utun.
45.	Ternorua........	Teranrua....	Terantolo	Rátua.
46.	Arúa......................	Oturúa.............	Otil	Rása.
49.	Fufu lu	Lufu lu	Lufu tol............	Uton.
50.	Yah mulu	Ya luh......... ....Ya tol		.Toon.
59.	....			.Datus.

INDEX.

INDEX TO VOLS. I. AND II.

ties of taking a house, 197; traders of, 200, 201; articles for exchange, 202; town of, 213; merchandise of, 213, 214; manners and customs, 214; various races of, 214, 215; absence of laws, 215; the genius of commerce at work, *ib.*; departure from, 218; map of, 219; trading at, 244, 245; second residence at, 267; its improved and animated appearance, 268 *et seq.*; cockfighting and football at, 269; cheapness of European articles of commerce, 271, 272; intemperance of the natives, 272; the author's recovery from a long illness, 275; mortality at, 278: funeral ceremonies at, *ib.*; active preparations for leaving, 279; extensive trade carried on at, 281.

Dodinga, village of, ii. 14; Portuguese fort at, 15.

Dogs, their voracity, ii. 259, 260.

Doleschall, Dr., in Amboyna, i. 458; his collection of flies and butterflies, 461.

Dorey, harbour and village of, ii. 304, 305; inhabitants of, 305, 306; house-building at, 307, 308; bird-shooting at, 309; the country round about, 311; the author's protracted sickness at, 316, 317; rudimental art among the people, 324; beetles and butterflies of, 326; numerous species of beetles at, 326, 327; expectations of disappointed, 328; departure from, 329.

Dorey vocabulary, ii. 475.

Doves at Malacca, i. 44.

Drusilla catops, ii. 199.

Duivenboden, Mr., known as the King of Ternate, ii. 2; his character, *ib.*

Durian and Mangusteen fruit. i. 83; and Durian, 116, 117; the Durian tree, 117; richness and excellency of, 118; dangerous when it falls from the trees, 119, 217.

Dutch, in Malacca, i. 42; in Java,

148; excellency of their colonial government (*see* Java); paternal despotism, 398-400; the cultivation system, 401; female labour, 403; their influence established in the Malay seas, ii. 7; their praiseworthy efforts to improve the Amboynese of the Malay Archipelago, 80.

Dutch mail steamer, life on board, i. 447.

Dyak house, i. 83, 84; Dyak mode of climbing a tree, 85, 86; Dyak dogs, 88; Dyak accounts of the Mias, 94, 95 (*see* Tabókan); agriculture, &c. 109, 111; houses, bridges, &c. 121-124; the character of the race in its relations to kindred ones, 137; higher in mental capacity than the Malays, 138; amusements of the young, *ib.*; moral character, 139; the Nile Dyaks never go to sea, *ib.*; head-hunting, *ib.*; truthfulness of, 139, 140; honesty, temperance, &c., *ib.*; checks of population, 141, 142; hard work of the women, 143; and idleness of the men, *ib.*; benefits arising from the government of Sir James Brooke, 144-146.

E.

Earl, Mr. George Windsor, his paper and pamphlet on the "Physical Geography of South-Eastern Asia and Australia," i. 13, 14.

Earthquakes at Ternate, ii. 9, 10.

Eclectus grandis, ii. 32

Elephants in Malacca, i. 52.

Elephomia, of New Guinea, different species, ii. 313-315.

Empugnan, a Malay village, i. 102 (*see* Tabókan), 103.

Entomological specimens, ii. 32.

Erythina, ii. 277.

Eucalypti, common tree of Timor, i. 12.

Eurystomus azureus, ii. 42.

Malay Archipelago, ii. 1 *et seq.*;
the author's views as to the races
of man in the, 439 *et seq.*; two
strongly contrasted races, the
Malays and the Papuans, 439 (*see*
Malays and Papuans); an indigenous race in the island of
Ceram, 449; tribes of the island
of Timor, *ib.*; the black woolly-
haired races of the Philippines
and the Malay peninsula, 451;
general view as to their origin and
affinities, 452, 453; the Polynesian races, 454, 455; on the
crania and the races of man in
the, 465–467.

Malay peninsula, non-volcanic, i. 11.

Malay race of Ternate, ii. 10.

Malay vocabulary, ii. 472.

Malays, a peculiarly interesting race,
found only in the Malay Archipelago, i. 1, 2; villages, &c., destroyed by volcanoes, 7–10; in
Singapore, 31; a Malay Governor
and house, 101 (*see* Gudong);
Malay villages, 195; of Batchian,
ii. 43; different from the Papuans,
178; contrast between the, 179;
contrast of character with the,
193; psychology of the, 207;
widely separated from the Papuans, 208; of the Malay Archipelago, 439; the most important
of the races, *ib.*; their physical
and mental characteristics, 440;
different tribes, *ib.*; the savage
Malays, 441; personal characteristics, *ib.*; impassive character
of, 442, 443; different accounts
of them, 444; on the crania and
languages of the, 466, 468.

Maleos, singular birds, in Celebes,
i. 413; description of the birds,
416; and eggs, 416–418; their
breeding-place, 475.

Mammalia, or warm-blooded quadrupeds of the Indo-Malay Islands,
i. 227; of the Timor group, 326;
of Celebes, 432; of the Moluccas,
ii. 138; of New Guinea, 429.

Mangusteen fruit (*see* Durian), in
Saráwak, i. 132, 217.

Manipa, island of, ii. 135.

Manowolko, the largest of the Goram
group, ii. 94; map of, 95; description of, 96; people and races
of, *ib.*; return to, 104.

Mansinam, island of, ii. 301, 304.

Manuel, a Portuguese bird-skinner,
engaged by the author, i. 242; his
philosophy, 250–253.

Mareh, island of, ii. 26, 27.

Maros river, i. 359; falls of the river,
367; precipices, 371; absence of
flowers, 373; drought followed by
a deluge of rain, 375; effects of,
376 (*see* Menado).

Marsupials in the Malay Archipelago, i. 13; of Celebes, 436;
of the Moluccas, ii. 142; of the
Papuan Islands, 428.

Massaratty vocabulary, ii. 473.

Matabello Islands, ii. 97, 98; dangers of the voyage to, 98, 99;
trade of, 100; cocoa-nut trees of,
ib.; villages of, 110; savage life of,
ib.; palm-wine one of the few
luxuries of, 102; wild fruits of,
103; strange ideas of the people
respecting the Russian war, 104;
their extravagant notions of the
Turks, *ib.*

Matabello vocabulary, ii. 475.

Mataram, capital of Lombock, i.
255, 256.

Mats and boxes of the Aru Islands,
ii. 251.

Mausoleum, ancient, in Java, i. 162,
163.

"Max Havelaar," story of the Dutch
auctions, &c. in the colonies,
i. 152.

Maykor, map of, ii. 219; river of,
286.

Megachile pluto, ii. 68.

Megamendong mountain, road over,
i. 176; a residence on, *ib.*; collections on, and in the neighbourhood of, 177–189.

Megapodidæ, the, a small family
of birds, peculiar to Australia,
and to surrounding islands, i.
243.

Megapodii of the Moluccas, ii. 147.

Pearl shells, the chief staple of the Aru trade, ii. 287.

Pelah, bad account of, ii. 29; journey to, 130.

Penrissen Mountains, at the head of the Saráwak river, i. 113.

Peters, Mr. of Awaiya, ii. 78.

Phalænopis grandiflora, ii. 189.

Pheasant, great Argus, country of, i. 51.

Pheasants, in Sumatra and Borneo, i. 167, 168.

Philippine Islands, i. 6; active and extinct volcanoes in, 9, 11; black woolly races of the, ii. 451.

Phosphoric light, rushing streams of, ii. 167.

Physical geography (*see* Archipelago).

Pieris, genus, ii. 83.

Pig, wild, i. 22, 433.

Pigs, their power of swimming, ii. 141.

Pigeons, fruit, i. 189; various, 244, 245; several species of, 293; of immense size, ii. 181; of New Guinea, 430

Pin, a strange novelty to the natives, ii. 134.

Pirates, on the Batchian coast, ii. 59; Sir J. Brooke's suppression of, on the coast of Borneo, *ib.*; on the coast of Aru, 210; attack the praus and murder the crews, 211, 212.

Pitcher plant, on Mount Ophir, i. 48; water in, 49; the plant in Borneo, 127, 216.

Pitta genus, ii. 136.

Pitta celebensis, ii. 136.

Pitta gigas, a beautiful bird of Gilolo, ii. 16.

Plants, on Mount Ophir (Ferns and Pitcher Plants, *see* both), i. 48, 49; rhododendrons, *ib.* ; zingiberaceous plants, 51 (*see* Durian and Bamboo) ; on Pangerango mountain, 181–185; geographical distribution of, ii. 293, 295 *et seq.*; distribution of, in New Guinea, 437.

Plough, a native, i. 353 ; ploughing, 353, 354.

Plumage of Birds of Paradise, changes of, ii. 398 *et seq.*

Polynesia, an area of subsidence, ii. 457.

Polynesian races, ii. 454, 455, 458; on the crania and languages of the, 467, 468.

Pomali, or "taboo," i. 306.

Poppa, map of, ii. 332; difficulties near the island of, 337 *et seq.*

Portuguese, in Singapore, from Malacca, i. 31; in Malacca, 40, 41; bad government of, in Timor, 307; expelled from Ternate by the Dutch, ii. 7; truly wonderful conquerors and colonizers, 192.

Pottery, carved tool for making, ii. 324.

Prau, native, of Macassar, ii. 160, 161; the crew, 163, 168, 169; captain and owner of the, 170; dangerous defects of the, 171, 172; comforts of the, 194.

Primula imperialis in Java, i. 183.

Productions, natural, contrasts of in the Malay Archipelago, i. 14; peculiarities of position in certain localities, 15–17; natural means of dispersal of, 16; a supposed case of natural dispersal, 25, 26; an exact parallel in the Malay Archipelago, 27, 28.

Ptilonopus pulchellus, ii. 354.

Ptilonopus superbus, and P. iogaster, ii. 54.

Ptiloris alberti, of N. Australia, ii. 417.

Pumbuckle chief, in Lombock, i. 259.

R.

Races, contrasts of, i. 29; two distinct, in the Archipelago, the Malays and Papuans, 29, 30, ii. 439 *et seq.* (*see* Malays and Papuans); opinions of Humboldt and Pritchard, i. 29; indigenous race in the island of Ceram, ii. 449 ; the Timorese, 449, 450; the black woolly-haired races of the Philippine and the Malay peninsula,

Therates labiata, ii. 191.

Thieving, trial and punishment for, ii. 216.

Thrushes, fruit, i. 21, 22; leaf, 22; ground, 245; beauty of, 246.

Tides, curious phenomena attending, ii. 371, 372.

Tidore, island of, i. 6; volcanic cone of, ii. 1; village of, 26; Sultan of, 357.

Tidore vocabulary, ii. 474.

Tiger, in Singapore, i. 36, 37; traps for, and dinners of, *ib.*; in Malacca, 52; tiger-cats in Borneo, 61; a tiger-hunt, 168, 169.

Timor, volcano in, i. 9; no forest in, 11; common trees of, 12; extent, &c. of, 288 (*see* Coupang and Delli); mountains of, 304, 305; value of the island, 308, 309; races of, ii. 449, 450; races of the islands west of, 450.

Timor group of islands, natural history of, i. 316; birds in, 318–322; fauna, 324, 325; mammalia, 326–329.

Timor vocabulary, ii. 475.

Tobacco, exchanged for insects, ii. 187.

Tobo, village of, ii. 89, 90.

Todiramphus diops, ii. 54.

Tomboro, volcano of, i. 7; great eruption of, *ib.*

Tomohon vocabulary, ii. 473.

Tomoré, colony from, ii. 43; people of East Celebes, 135.

Tomoré vocabulary, ii. 473.

Tondano, village of, in Celebes, i. 349; waterfall at, 395, 396; from Tondano to Kakos (which see).

Towers, Mr., an Englishman resident in Menado, i. 378.

Trade, in Lombock, i. 274, 275 (*see* Census); the magic that keeps all at peace, ii. 215, 216; very considerable at Dobbo, 281, 282.

Tree, large, at Modjokerto, in Java, i. 156.

Tree ferns, of immense size, ii. 209.

Tree-kangaroos, ii. 319, 320.

Tricondyla aptera, ii. 191.

Tripang, produce of, ii. 114.

Trogons, i. 22, 24.

Tropidorhynchi of the Moluccas, strong and active birds, ii. 152, 153.

Tropidorhynchus bouruensis, ii. 151; T. subcornutus, 152.

Tropidorhynchus fuscicapillus, ii. 21.

Turks, extravagant ideas of, entertained by the Malay Archipelagans, ii. 104.

U.

Untowan mountains, i. 114.

Uta, island of, ii. 99, 100.

V.

Vaiqueno vocabulary, ii. 475.

Vanda lowii, the plant, in Borneo, i. 128.

Vegetation, contrasts of, in the Malay Archipelago, i. 11; European in Java, 184, 185; in Timor, 310, 311; in Celebes, 389, 390.

Villages of the Sumatran Malays, i. 195; pretty villages in Celebes, 383.

Violets, &c. in Java, i. 184.

Viverra tangalunga, ii. 139.

Viverra zebetha, ii. 55.

Vocabularies, list of, collected, ii. 472.

Volcanic and non-volcanic islands, contrasts of, i. 6; volcanic belt, course and extent of, 6, 10, 24.

Volcanoes, i. 6–10; mud, 407; view of the volcano at Bali, 450-452; in Amboyna, 460, 461; of the Malay Islands, ii. 27; elevation and depression of the land arising from, 30.

Vorkai river, ii. 286.

Vorkay, map of, ii. 219.

W.

Wahai, village of, ii. 91; arrival at, 123.

Wahai vocabulary, ii. 475.

THE END.

LONDON: R. CLAY, SONS, AND TAYLOR, PRINTERS.

Printed in the United States
By Bookmasters